T0270914

MATRIX POSITIVITY

Matrix positivity is a central topic in matrix theory: properties that generalize the notion of positivity to matrices arose from a large variety of applications, and many have also taken on notable theoretical significance, either because they are natural or unifying or because they support strong implications. This is the first book to provide a comprehensive and up-to-date reference of important material on matrix positivity classes, their properties, and their relations. The matrix classes emphasized in this book include the classes of semipositive matrices, P-matrices, inverse M-matrices, and copositive matrices.

This self-contained reference will be useful to a large variety of mathematicians, engineers, and social scientists, as well as graduate students. The generalizations of positivity and the connections observed provide a unique perspective, along with theoretical insight into applications and future challenges. Direct applications can be found in data analysis, differential equations, mathematical programming, computational complexity, models of the economy, population biology, dynamical systems, and control theory.

CHARLES R. JOHNSON is Class of 1961 Professor of Mathematics at College of William and Mary. He received his Ph.D. from The California Institute of Technology in 1972. Professor Johnson has published nearly 500 papers and 12 books and received several prizes.

RONALD L. SMITH is Professor Emeritus at the University of Tennessee, Chattanooga. He received his Ph.D. in Mathematics at Auburn University. He received the Thomson-Reuters Original List of Highly Cited Researchers Award. The conference "Recent Advances in Linear Algebra and Graph Theory" was held in his honor.

MICHAEL J. TSATSOMEROS is Professor of Mathematics at Washington State University. He received his Ph.D. in Mathematics at the University of Connecticut in 1990. He has served on the Board of Directors and Advisory and Journal Committees of the International Linear Algebra Society. He is co-Editor-in-Chief of the *Electronic Journal of Linear Algebra* and Associate Editor of *Linear Algebra and Its Applications*.

CAMBRIDGE TRACTS IN MATHEMATICS

A complete list of books in the series can be found at www.cambridge.org/mathematics. Recent titles include the following:

187. Convexity: An Analytic Viewpoint. By B. SIMON
188. Modern Approaches to the Invariant Subspace Problem. By I. CHALENDAR and J. R. PARTINGTON
189. Nonlinear Perron-Frobenius Theory. By B. LEMMENS and R. NUSSBAUM
190. Jordan Structures in Geometry and Analysis. By C.-H. CHU
191. Malliavin Calculus for Lévy Processes and Infinite-Dimensional Brownian Motion. By H. OSSWALD
192. Normal Approximations with Malliavin Calculus. By I. NOURDIN and G. PECCATI
193. Distribution Modulo One and Diophantine Approximation. By Y. BUGEAUD
194. Mathematics of Two-Dimensional Turbulence. By S. KUKSIN and A. SHIRIKYAN
195. A Universal Construction for Groups Acting Freely on Real Trees. By I. CHISWELL and T. MÜLLER
196. The Theory of Hardy's Z-Function. By A. IVIĆ
197. Induced Representations of Locally Compact Groups. By E. KANIUTH and K. F. TAYLOR
198. Topics in Critical Point Theory. By K. PERERA and M. SCHECHTER
199. Combinatorics of Minuscule Representations. By R. M. GREEN
200. Singularities of the Minimal Model Program. By J. KOLLÁR
201. Coherence in Three-Dimensional Category Theory. By N. GURSKI
202. Canonical Ramsey Theory on Polish Spaces. By V. KANOVEI, M. SABOK, and J. ZAPLETAL
203. A Primer on the Dirichlet Space. By O. EL-FALLAH, K. KELLAY, J. MASHREGHI, and T. RANSFORD
204. Group Cohomology and Algebraic Cycles. By B. TOTARO
205. Ridge Functions. By A. PINKUS
206. Probability on Real Lie Algebras. By U. FRANZ and N. PRIVAULT
207. Auxiliary Polynomials in Number Theory. By D. MASSER
208. Representations of Elementary Abelian p-Groups and Vector Bundles. By D. J. BENSON
209. Non-homogeneous Random Walks. By M. MENSHIKOV, S. POPOV and A. WADE
210. Fourier Integrals in Classical Analysis (Second Edition). By C. D. SOGGE
211. Eigenvalues, Multiplicities and Graphs. By C. R. JOHNSON and C. M. SAIAGO
212. Applications of Diophantine Approximation to Integral Points and Transcendence. By P. CORVAJA and U. ZANNIER
213. Variations on a Theme of Borel. By S. WEINBERGER
214. The Mathieu Groups. By A. A. IVANOV
215. Slenderness I: Foundations. By R. DIMITRIC
216. Justification Logic. By S. ARTEMOV and M. FITTING
217. Defocusing Nonlinear Schrödinger Equations. By B. DODSON
218. The Random Matrix Theory of the Classical Compact Groups. By E. S. MECKES
219. Operator Analysis. By J. AGLER, J. E. MCCARTHY, and N. J. YOUNG
220. Lectures on Contact 3-Manifolds, Holomorphic Curves and Intersection Theory. By C. WENDL
221. Matrix Positivity. By C. R. JOHNSON, R. L. SMITH and M. J. TSATSOMEROS

Matrix Positivity

CHARLES R. JOHNSON
College of William and Mary

RONALD L. SMITH
University of Tennessee at Chattanooga

MICHAEL J. TSATSOMEROS
Washington State University

CAMBRIDGE
UNIVERSITY PRESS

University Printing House, Cambridge CB2 8BS, United Kingdom

One Liberty Plaza, 20th Floor, New York, NY 10006, USA

477 Williamstown Road, Port Melbourne, VIC 3207, Australia

314–321, 3rd Floor, Plot 3, Splendor Forum, Jasola District Centre,
New Delhi – 110025, India

79 Anson Road, #06–04/06, Singapore 079906

Cambridge University Press is part of the University of Cambridge.

It furthers the University's mission by disseminating knowledge in the pursuit of education, learning, and research at the highest international levels of excellence.

www.cambridge.org
Information on this title: www.cambridge.org/9781108478717
DOI: 10.1017/9781108778619

First published 2020

A catalogue record for this publication is available from the British Library.

Library of Congress Cataloging-in-Publication Data

ISBN 978-1-108-47871-7 Hardback

To the many students who have helped me with mathematics through my REU program at William and Mary. They have made it fun for me as well as enhanced greatly what I have been able to do.

Charles R. Johnson

To my parents and siblings; to Linda, my wife and best friend, whose help on this book was invaluable to me; and to Kevin, Brad, Reid, and Sarah, our children who gave us six wonderful grandchildren.

Ronald L. Smith

To Μόσχα, Katerina, and Joanne. Your inspiration and support make it all possible.

Michael J. Tsatsomeros

Contents

Preface

Properties that generalize the notion of positivity (of scalars) to matrices have arisen from a large variety of applications. Many have also taken on notable theoretical significance, either because they are natural or unifying or because they support strong implications. All three authors have written extensively on matricial generalizations of positivity, over a long period of time. Some of these generalizations are already the subject of one or more books, sometimes with excellent and modern treatments. But, for a variety of reasons, others are not yet treated comprehensively in readily available form. We feel that a good reference for these will be a useful contribution. This is the purpose of the present work.

After a review of relevant background in Chapter 1, we discuss the most prominent generalizations of positivity in Chapter 2. The containment and other relationships among these are given, as well as useful book references for some. It is shown that all of them are contained among the "semipositive" matrices. There are, of course, variations upon these prominent generalizations too, and there are often weak and strong versions of the generalizations.

In subsequent chapters, particular generalizations of positivity are studied in detail, when we feel that we can make significant and novel contributions. These include semipositive matrices in Chapter 3, P-matrices in Chapter 4, Inverse M-matrices in Chapter 5, and copositive matrices in Chapter 6. Then we conclude with a lengthy list of references for these areas.

We thank Wenxuan Ding, Yuqiao Li, Yueqiao Zhang, and Megan Wendler for their help in the proofreading and preparation of parts of the manuscript.

Symbols

- $\mathbb{R}$, $\mathbb{C}$ Fields of real, complex numbers
- $\mathbb{R}^n$, $\mathbb{C}^n$ Column n-vectors of real, complex numbers
- $\mathbb{R}^n_+$ Nonnegative orthant in $\mathbb{R}^n$ (all n-vectors of nonnegative numbers)
- $M_{m,n}(\mathbb{F})$ The m-by-n matrices over field $\mathbb{F}$; skipping $\mathbb{F}$ means $\mathbb{F} = \mathbb{C}$
- $M_n(\mathbb{F}) = M_{n,n}(\mathbb{F}),\ M_n = M_n(\mathbb{C}) = M_{n,n}(\mathbb{C})$
- $M_n(\{-1,0,1\})$ The n-by-n matrices with entries in $\{-1,0,1\}$
- $M_n(\{-1,1\})$ The n-by-n matrices with entries in $\{-1,1\}$
- X^T, X^* Transpose and conjugate transpose of a complex array X
- $X^\dagger$ Moore–Penrose inverse of $X \in M_{m,n}(\mathbb{C})$
- $X^\#$ Group inverse of $X \in M_{m,n}(\mathbb{C})$
- $X > Y$ ($X \geq Y$) Real arrays X, Y, every entry of $X - Y$ is positive (nonnegative)
- $[X, Y]$ Matrix interval ($Y \geq X$), all real matrices Z such that $X \leq Z \leq Y$
- $\langle n \rangle = \{1, 2, \ldots, n\}$.
- $A \circ B$ Hadamard (entry-wise) product of $A, B \in M_{m,n}(\mathbb{F})$
- $A^{(k)}$ k-th Hadamard power $A \circ A \circ \cdots \circ A$, $A \in M_{m,n}(\mathbb{F})$
- index(A) Index of A
- Δ_n Unit simplex in $\mathbb{R}^n$
- J All ones square matrix
- e All ones column vector
- Tr(A) Trace of $A \in M_n(\mathbb{F})$
- $Q(x) = x^T Ax$ Associated quadratic form, $A \in M_n(\mathbb{C})$, $x \in \mathbb{C}^n$
- adj A Adjoint of $A \in M_{m,n}(\mathbb{C})$
- $R(A)$ Range of $A \in M_{m,n}(\mathbb{C})$
- rank(A) Rank of $A \in M_{m,n}(\mathbb{C})$
- Null(A) (Right) null space of $A \in M_{m,n}(\mathbb{C})$
- nullity(A) Dimension of Null(A)
- $\sigma(A)$ Spectrum (eigenvalues) of $A \in M_n(\mathbb{C})$

- $\rho(A) = \max\{|\lambda| : \lambda \in \sigma(A)\}$ Spectral radius of $A \in M_n(\mathbb{C})$
- $q(A)$ Positive eigenvalue of minimum modulus of an M-matrix A
- $\mathrm{diag}(d_1, d_2, \ldots, d_n)$ The n-by-n diagonal matrix with diagonal entries $d_1, d_2 \ldots, d_n$
- $|\alpha|$ Cardinality of $\alpha \subseteq \langle n \rangle$
- $\alpha^c = \langle n \rangle \setminus \alpha$ Complement of $\alpha \subseteq \langle n \rangle$ in $\langle n \rangle$
- $A[\alpha, \beta]$ Submatrix of A lying in rows $\alpha \subseteq \langle m \rangle$ and columns $\beta \subseteq \langle n \rangle$
- $A[\alpha] = A[\alpha, \alpha]$ Principal submatrix of A lying in rows $\alpha \subseteq \langle n \rangle$
- $A(\alpha, \beta) = A[\alpha^c, \beta^c]$
- $A(i) = A(\{i\})$
- $A/A[\alpha]$ Schur complement of $A[\alpha]$ in $A \in M_n(\mathbb{F})$, $\alpha \subseteq \langle n \rangle$
- $\mathrm{ppt}\,(A, \alpha)$ Principal pivot transform of $A \in M_n(\mathbb{C})$ relative to $\alpha \subseteq \langle n \rangle$
- F_A Cayley transform of $A \in M_n(\mathbb{C})$
- $\mathcal{P}(A/A[\alpha])$ Perron complement of $A[\alpha]$ in $A \in M_n(\mathbb{F})$, $\alpha \subseteq \langle n \rangle$
- $S_+(A) = \{x \in \mathbb{R}^n : x \geq 0 \text{ and } Ax > 0\}$, $A \in M_{m,n}(\mathbb{R})$
- $R_+(A) = A\, S_+(A)$, $A \in M_{m,n}(\mathbb{R})$
- $K_+(A) = \{x \in \mathbb{R}^n : x \geq 0 \text{ and } Ax \geq 0\}$, $A \in M_{m,n}(\mathbb{R})$
- $A \oslash B = \begin{bmatrix} A \\ B \end{bmatrix} \in M_{m+p,n}(\mathbb{R}), \quad A \in M_{m,n}(\mathbb{R})$ and $B \in M_{p,n}(\mathbb{R})$
- $A \ominus B = [A\ B] \in \mathrm{M}_{m,n+q}, \quad A \in M_{m,n}(\mathbb{R})$ and $B \in M_{m,q}(\mathbb{R})$
- $\mathcal{M}(A)$ Comparison matrix of $A \in M_n(\mathbb{C})$
- $D(A)$ Directed graph of $A \in M_n(\mathbb{C})$
- $S(A)$ Signed directed graph of $A \in M_n(\mathbb{R})$
- $\mathrm{LCP}\,(q, M)$ The Linear Complementarity Problem, $M \in M_n(\mathbb{R})$, $q \in \mathbb{R}^n$
- $m(A)$ Measure of irreducibility of $A \in M_n(\mathbb{C})$
- $U(A)$ Upper path product bound for $A \in M_n(\mathbb{C})$
- $I(A, B)$ Interval from $A \in M_n(\mathbb{R})$ to $B \in M_n(\mathbb{R})$
- $V(A, B)$ Vertex matrices derived from $A, B \in M_n(\mathbb{R})$

Matrix Classes

- **C** (**C**$_n$) Copositive matrices (n-by-n)
- **CP** Completely positive matrices
- D_n n-by-n invertible diagonal matrices
- D_n^+ n-by-n positive diagonal matrices in D_n
- **DN** Doubly nonnegative matrices
- **DP** Doubly positive matrices
- **EIM** Eventually inverse M-matrices
- **EP** (Entry-wise) positive matrices
- **IM** Inverse M-matrices

- $\mathbf{IM}^D$ Dual of **IM**
- **IS** Identically signed class of matrices
- **LSP** Left semipositive matrices
- M_n^k $A \in M_n(\mathbb{R})$ with nonzero diagonal entries and length of longest cycle in $D(A) \leq k$
- **M** M-matrices
- **MSP** Minimally semipositive matrices
- **N** (Entry-wise) Nonnegative matrices
- $\mathbf{N}^+$ Nonnegative matrices with positive main diagonal
- **ND** Negative definite matrices
- **NSD** Negative semidefinite matrices
- P_n Group of n-by-n permutation matrices
- P_n^k Set of real P-matrices in M_n^k
- **P** P-matrices
- $\mathbf{P}_\mathrm{M}$ Matrices all of whose powers are P-matrices
- $\mathbf{P}_0$ P_0-matrices
- **PD** Positive definite matrices
- **PSD** Positive semidefinite matrices
- **PSPP** Purely strict path product matrices
- **PP** Path product matrices
- **RSP** Redundantly semipositive matrices
- $\mathbf{S}_n = \mathbf{S}_n(\mathbb{R})$ Set of symmetric matrices in $M_n(\mathbb{R})$
- S_n^k Set of $A \in M_n(\mathbb{R})$ all of whose cycles in $S(-A)$ are signed negatively
- **SC** ($\mathbf{SC}_n$) Strictly copositive matrices (n-by-n)
- **SIM** Symmetric inverse M-matrices
- **sN** Symmetric nonnegative matrices
- **SN** Seminegative matrices
- **SNN** Seminonnegative matrices
- **SNP** Seminonpositive matrices
- **SP** Semipositive matrices
- **SPP** Strict path product matrices
- **SZ** Semizero matrices
- **TN** Totally nonnegative matrices
- **TP** Totally positive matrices
- **TSPP** Totally strict path product matrices
- **Z** Z-matrices

1

Background

1.1 Purpose

Here, we review some well-known mathematical concepts that are helpful in developing the ideas of this work. These are primarily in the areas of matrix analysis, convexity, and an important theorem of the alternative (linear inequalities). In the latter two cases, we mention all that is needed. In the case of matrices, see the general reference [HJ13], or a good elementary linear algebra book, for facts or notation we use without further explanation.

1.2 Matrices

1.2.1 Matrix and Vector Notation

We use $\mathbb{R}^n$ ($\mathbb{C}^n$) to denote the set of all n-component real (complex) vectors, thought of as columns, and $M_{m,n}(\mathbb{F})$ to denote the m-by-n matrices over a general field $\mathbb{F}$. Skipping the field means $\mathbb{F} = \mathbb{C}$ and $M_{n,n}(\mathbb{F})$ is abbreviated to $M_n(\mathbb{F})$.

Inequalities, such as $>$, $\geq$, are to be interpreted entry-wise, so that $x > 0$ means that all entries of a vector x are positive. If all entries of x are nonnegative, but not all zero, we write $x \geq 0, \neq 0$.

The **transpose** of $A = [a_{ij}] \in M_{m,n}$ is denoted by A^T and the **conjugate transpose** (or **Hermitian adjoint**) by A^*. The **Hermitian** and **skew-Hermitian** parts of A are, respectively, denoted by

$$H(A) = \frac{A + A^*}{2} \quad \text{and} \quad S(A) = \frac{A - A^*}{2}.$$

The **spectrum**, or set of eigenvalues, of $A \in M_n$ is denoted by $\sigma(A)$ and the **spectral radius** by $\rho(A) = \max_{\lambda \in \sigma(A)} |\lambda|$.

Submatrices play an important role in analyzing matrix structure. For $\langle m\rangle = \{1,2,\dots,m\}$ and $\langle n\rangle = \{1,2,\dots,n\}$, denote by $A[\alpha,\beta]$ the **submatrix** of $A \in M_{m,n}(\mathbb{F})$ lying in rows $\alpha \subseteq \langle m\rangle$ and columns $\beta \subseteq \langle n\rangle$. If $m = n$ and $\alpha = \beta$, we abbreviate $A[\alpha,\alpha]$ to $A[\alpha]$ and refer to it as a **principal submatrix** of A; we call $A[\alpha]$ a **proper principal submatrix** if α is a proper subset of $\langle n\rangle$. A submatrix of $A \in M_n$ of the form $A[\langle k\rangle]$ for some $k \le n$ is called a **leading principal submatrix**. A submatrix may also be indicated by deletion of row and column indices, and for this round brackets are used. For example, $A(\alpha,\beta) = A[\alpha^c,\beta^c]$, in which the complementation is relative to $\langle m\rangle$ and $\langle n\rangle$, respectively. If an index set is a singleton i, we abbreviate $A(\{i\})$, for example, to $A(i)$ in case $m = n$.

Given a square submatrix $A[\alpha,\beta]$ of $A \in M_{m,n}$, we refer to $\det(A[\alpha,\beta])$ as a **minor** of A, or as a **principal minor** if $\alpha = \beta$. The determinant of a leading principal submatrix is referred to as a **leading principal minor**. By convention, if $\alpha = \emptyset$, then $\det(A[\alpha]) = 1$.

1.2.2 Gershgorin's Theorem

For $A = [a_{ij}] \in M_n(\mathbb{C})$, for each $i = 1,2,\dots,n$, define

$$R_i'(A) = \sum_{j=1, j\neq i}^{n} |a_{ij}| \quad \text{and the discs} \quad \Gamma_i(A) = \{z \in \mathbb{C} : |z - a_{ii}| \le R_i'(A)\}.$$

Gershgorin's Theorem then says that $\sigma(A) \subseteq \cup_{i=1}^{n} \Gamma_i(A)$. This has the implication that a **diagonally dominant matrix** $A \in M_n(\mathbb{C})$ (i.e., $|a_{ii}| > R_i'(A)$, $i = 1,2,\dots,n$) has nonzero determinant. In particular, the determinant is of the same sign as the product of the diagonal entries, in the case of real matrices.

1.2.3 Perron's Theorem

If $A \in M_n(\mathbb{R})$, $A > 0$, then very strong spectral properties follow as initially observed by Perron (**Perron's Theorem**). These include

- $\rho(A) \in \sigma(A)$; (1.2.1)
- the multiplicity of $\rho(A)$ is one; (1.2.2)
- $\lambda \in \sigma(A)$, $|\lambda| = \rho(A) \implies \lambda = \rho(A)$; (1.2.3)
- there is a right (left) eigenvector of A with all entries positive; (1.2.4)
- $0 < B \le A$, $B \neq A$ implies $\rho(B) < \rho(A)$; (1.2.5)
- No eigenvector (right or left) of A has all nonnegative entries besides those associated with $\rho(A)$. (1.2.6)

There are slightly weaker statements for various sorts of nonnegative matrices; see [HJ91, chapter 8].

1.2.4 Schur Complements

If

$$A = \begin{bmatrix} A_{11} & A_{12} \\ A_{21} & A_{22} \end{bmatrix} \in M_n$$

and A_{22} is square and invertible, the **Schur complement** (of A_{22} in A) is

$$A/A_{22} = A_{11} - A_{12}A_{22}^{-1}A_{21}.$$

More generally, if $\alpha \subseteq \langle n \rangle$ is an index set, then

$$A/A[\alpha] = A(\alpha) - A[\alpha^c, \alpha]A[\alpha]^{-1}A[\alpha, \alpha^c]$$

is the Schur complement of $A[\alpha]$ in A if $A[\alpha]$ is invertible. Schur complements enjoy many nice properties, such as

$$\det A = \det A[\alpha] \, \det(A/A[\alpha]),$$

which motivates the notation. Reference [Zha05] is a good reference on Schur complements and contains a detailed discussion of Schur complements in positivity classes.

1.3 Convexity

1.3.1 Convex Sets in $\mathbb{R}^n$ and $M_{m,n}(\mathbb{R})$

A **convex combination** of a collection of elements of $\mathbb{R}^n$ is a linear combination whose coefficients are nonnegative and sum to 1. A subset S of $\mathbb{R}^n$ or $M_{m,n}(\mathbb{R})$ is **convex** if it is closed under convex combinations. It suffices to know closure for pairs of elements, and geometrically this means that a set is convex if the line segment joining any two elements lies in the set.

An **extreme point** of a convex set is one that cannot be written as a convex combination of two distinct points in the set. The **generators** of a convex set are a minimal set of elements from which any element of the set may be written as a convex combination. For finite dimensions (our case), the extreme points are the generators. If there are finitely many, the convex set is called **polyhedral**.

A (convex) **cone** is just a convex set that is closed under linear combination whose coefficients are nonnegative (no constraint on the sum). A cone may also be finitely generated (polyhedral cross section) or not. The **dual** of a cone $\mathcal{C}$ is just the set

$$\mathcal{C}^D = \{x\colon x \cdot y \geq 0, \quad \text{whenever } x \in \mathcal{C}\},$$

in which $\cdot$ denotes an inner product defined on the underlying space.

The **convex hull** of a collection of points is just the set of all possible convex combinations. This is a convex set, and any convex set is the convex hull of its generators.

1.3.2 Helly's Theorem

The intersection of finitely many convex sets is again a convex set (possibly empty). Given a collection of convex sets (possibly infinite) in a d dimensional space, the intersection of all of them is nonempty if and only if the intersection of any $d + 1$ of them is nonempty. This is known as **Helly's Theorem** [Hel1923] and can provide a powerful tool to show the existence of a solution to a system of equations.

1.3.3 Hyperplanes and Separation of Convex Sets

In an inner product space of dimension d, such as $\mathbb{R}^d$, a special kind of convex set is a **half-space**

$$H_a = \{x\colon a \cdot x \geq 0\},$$

in which $a \neq 0$ is a fixed element of the inner product space. Any intersection of half-spaces is a convex set, and any closed convex set is an intersection of half-spaces. The complement of a half-space,

$$H_a^c = \{x\colon a \cdot x < 0\},$$

is an open half-space. The set

$$\{x\colon a \cdot x = 0\}$$

is a $(d - 1)$-dimensional **hyperplane** that separates the two half-spaces H_a and H_a^c. If two convex sets S_1, S_2 in a d-dimensional space do not intersect, then they may be separated by a $(d - 1)$-dimensional hyperplane, so that $S_1 \subseteq H_a$ and $S_2 \subseteq H_{-a}$. If S_1 and S_2 are both closed or both open, we may use nonintersecting open half-spaces: $S_1 \subseteq H_a^c$ and $S_2 \subseteq H_{-a}^c$. If S_1 and S_2 do intersect but only in a subset of a hyperplane, then such a hyperplane may be used to separate them: $S_1 \subseteq H_a$ and $S_2 \subseteq H_{-a}$.

For further background on convex sets, see the general reference [Rock97].

1.4 Theorem of the Alternative

In the theory of linear inequalities (or optimization), there is a variety of statements saying that exactly one of two systems of linear inequalities has a solution. Such statements, and there are many variations, are called **Theorems of the Alternative** (and there are such theorems in even more general contexts). Such statements can be a powerful tool for showing that one system of inequalities has a solution, by ruling out the other; however, they generally are not able to provide any particular solution. The book [Man69] provides nice discussion of several theorems of the alternative and relations among them.

A particular version of the theorem of the alternative that is especially useful for us is the following:

Theorem 1.4.1 Let $A \in M_{m,n}(\mathbb{R})$. Then, either

(i) there is an $x \in \mathbb{R}^n$, $x \geq 0$, such that $Ax > 0$

or

(ii) there is a $y \in \mathbb{R}^m$, $y \geq 0$, $y \neq 0$, such that $y^T A \leq 0$,

but not both.

The "not both" portion is clear, as by (i), $y^T Ax > 0$ and because of (ii), $y^T Ax \leq 0$. In the event that $m = n$ and A is symmetric, we have

Corollary 1.4.2 Let $A \in M_n(\mathbb{R})$ and $A^T = A$. Then, either

(i) there is an $x \in \mathbb{R}^n$, $x \geq 0$, such that $Ax > 0$

or

(ii) there is a $y \in \mathbb{R}^m$, $y \geq 0$, $y \neq 0$, such that $Ay \leq 0$,

but not both.

2

Positivity Classes

2.1 Introduction

There are a remarkable number of ways that the notion of positivity for scalars has been generalized to matrices. Often, these represent the differing needs of applications, or differing natural aspects of the classical notion. Typically, the various ways involve the entries, transformational characteristics, the minors, the quadratic form, and combinations thereof. Our purpose here is to identify each of the generalizations and some of their basic characteristics. Usually there are natural variations that we also identify and relate. Several of these generalizations have been treated, in some depth, in book or survey form elsewhere; if so, we give some of the most prominent or accessible references. Our purpose in this work is to then treat in subsequent chapters those generalizations for which there seems not yet to be sufficiently general treatment in one place.

The order in which we give the generalizations is roughly grouped by type. We then summarize the containments among the positivity classes; one of them includes all the others.

2.2 (Entry-wise) Positive Matrices (EP)

One of the most natural generalizations of a positive number is reflected in the entries of a matrix. In general, adjectives, such as positive and nonnegative, refer to the entries of a matrix. The ways in which n-by-n positive matrices generalize the notion of a positive number are indicated in Perron's theorem; see Section 1.2.3.

Careful treatments of the theory of positive matrices may be found in several sources, such as [HJ13]. There are many further facts. Because of Frobenius's work on nonnegative matrices, the general theory is referred to as **Perron–Frobenius theory**.

There are important variants on positive matrices. For their closure, the nonnegative matrices, the spectral radius is still an eigenvalue, but all the other conclusions must be weakened somewhat. For nonnegative and **irreducible matrices** $\left(P^T AP \neq \begin{bmatrix} A_{11} & A_{12} \\ 0 & A_{22} \end{bmatrix}\right.$, with A_{11} and A_{22} square and nonempty for any permutation $\left.P\right)$, only property (1.2.3) must be weakened to allow other eigenvalues on the spectral circle. And, if some power, A^q, is positive, the stronger conclusions remain valid. The following are simple examples that illustrate what might occur.

$\begin{bmatrix} 1 & 1 \\ 1 & 1 \end{bmatrix}$ The Perron root (spectral radius) is strictly dominant when the matrix is entry-wise positive.

$\begin{bmatrix} 1 & 1 \\ 1 & 0 \end{bmatrix}$ It remains so if the matrix is not positive but some power is positive (primitive matrix).

$\begin{bmatrix} 0 & 1 \\ 1 & 0 \end{bmatrix}$ The Perron root remains of multiplicity 1, but there may be ties for spectral radius when the matrix is irreducible but not primitive.

$\begin{bmatrix} 1 & 0 \\ 0 & 1 \end{bmatrix}$, $\begin{bmatrix} 1 & 1 \\ 0 & 1 \end{bmatrix}$ The Perron root may be multiple and geometrically so, or not, when the matrix is reducible.

$\begin{bmatrix} 1 & 0 \\ 0 & 0 \end{bmatrix}$, $\begin{bmatrix} 1 & 0 \\ 1 & 0 \end{bmatrix}$ Or the Perron root may still have multiplicity 1, even when the matrix is reducible.

$\begin{bmatrix} 0 & 1 \\ 0 & 0 \end{bmatrix}$ The Perron root may be 0.

The **nonnegative orthant** in $\mathbb{R}^n$ is the closed cone that contains all entry-wise nonnegative vectors and is denoted by $\mathbb{R}^n_+$. The n-by-n **nonnegative matrices** ($A \geq 0$) are simply those that map the nonnegative orthant in $\mathbb{R}^n$ into itself ($A\mathbb{R}^n_+ \subseteq \mathbb{R}^n_+$). The n-by-n **positive matrices** ($A > 0$, or $A \in$ **EP** that stands for entry-wise positive) are those that map the nonzero elements of $\mathbb{R}^n_+$ into the interior of this cone. Matrices that map other cones of $\mathbb{R}^n$ with special structure into themselves emulate the spectral structure of nonnegative matrices, and this has been studied from several points of view in some detail.

Also studied have been matrices with some negative entries that still enjoy some or all of the above Perron conclusions or their Frobenius weakenings: $A \in M_n(\mathbb{R})$ is **eventually nonnegative (positive)** if there is an integer k such that $A^p \geq 0$ (> 0) for all $p \geq k$.

2.3 M-Matrices (M)

A matrix $A = [a_{ij}] \in M_n(\mathbb{R})$ is called a **Z-matrix** ($A \in$ **Z**) if $a_{ij} \leq 0$ for all $i \neq j$. Thus, a Z-matrix A may be written $A = \alpha I - P$ in which P is a nonnegative matrix. If $\alpha > \rho(P)$, then A is called a (non-singular) **M-matrix** ($A \in \mathbf{M}$). In several sources ([BP94, HJ91, FP62, NP79, NP80]) there are long lists of rather diverse-appearing conditions that are equivalent to A being an M-matrix, provided A is a Z-matrix. A modest list is the following:

1. A is **positive stable**, i.e., all eigenvalues of A are in the open right half-plane.
2. The leading principal minors of A are positive;
3. All principal minors of A are positive, i.e., A is a **P-matrix**;
4. A^{-1} exists and is a nonnegative matrix;
5. A is a semipositive matrix;
6. There exists a positive diagonal matrix D such that AD is row diagonally dominant;
7. There exist positive diagonal matrices D and E such that EAD is row and column diagonally dominant;
8. For each $k = 1, \ldots, n$, the sum of the k-by-k principal minors of A is positive;
9. A has an L–U factorization in which L and U have positive diagonal entries.

   In addition, M-matrices are closed under positive scalar multiplication, extraction of principal submatrices, and of Schur complements, and the so-called **Fan product** (which is the entry-wise or Hadamard product, except that the off-diagonal signs are retained). They are not closed under either addition or matrix multiplication.

   M-matrices, $A = [a_{ij}] \in M_n(\mathbb{R})$, also satisfy classical **determinantal inequalities**, such as

10. **Hadamard's inequality**: $\det A \leq \prod_{i=1}^{n} a_{ii}$;
11. **Fischer's inequality**: $\det A \leq \det A[\alpha] \det A[\alpha^c]$, $\alpha \subseteq \{1, \ldots, n\}$;
12. **Koteljanskii's inequality**: $\det A[\alpha \cup \beta] \det A[\alpha \cap \beta] \leq \det A[\alpha] \det A[\beta]$, $\alpha, \beta \subseteq \{1, \ldots, n\}$.

A complete description of all such principal minor inequalities is given in [Joh98].

M-matrices are not only positive stable, but, among Z-matrices, when all real eigenvalues are in the right half-plane, all eigenvalues are as well. They also have **positive diagonal Lyapunov solutions**, i.e., there is a positive diagonal matrix D such that $DA + A^T D$ is positive definite.

If $A = I - P$ is an irreducible M-matrix and we assume that $\rho(P) < 1$, then A is invertible and $A^{-1} = I + P + P^2 + \cdots$ Because P is irreducible, A^{-1} is positive. If A had been reducible, A^{-1} would be nonnegative. A matrix is **inverse M (IM)** if it is the inverse of an M-matrix, or if it is a nonnegative invertible matrix whose inverse is a Z-matrix. Among nonnegative matrices, the inverse M-matrices have a great deal of structure, and Chapter 5 is devoted to developing that structure for the first time in one place.

2.4 Totally Positive Matrices (TP)

A much stronger positivity requirement than that a matrix be positive is that all of its minors be positive. Such a matrix is called **totally positive** (**TP**). Such matrices arise in a remarkable variety of ways, and, of course, they also have very strong properties. The eigenvalues are positive and distinct, and the eigenvectors are highly structured in terms of the signs of their entries relative to the order in which the eigenvalue lies relative to the other eigenvalues. They always have an L–U factorization in which every minor of L and U is positive, unless it is identically 0. The determinantal inequalities of Hadamard, Fischer, and Koteljanskii (in Section 2.3) are satisfied, as well as many different ones. Transformationally, if A is **TP**, Ax cannot have more sign changes than x, and, of course, the **TP** matrices are closed under matrix multiplication. Though the definition requires many minors to be positive, because of Sylvester's determinantal identity, relatively few need be checked; the **contiguous minors** (both index sets are consecutive) suffice, and even the **initial minors** (those contiguous minors of which at least one index set begins with index 1) suffice. The initial minors are as numerous as the entries.

There are several comprehensive sources available for **TP** (and related) matrices, including the most recent book [FJ11]. Prior sources include the book [GK1935] and the survey [And80].

There are a number of natural variants on **TP** matrices. A **totally nonnegative** matrix, **TN**, is one in which all minors are nonnegative. **TN** is the topological closure of **TP**, and the properties are generally weaker. Additional variants are $\mathbf{TP}_k$ and $\mathbf{TN}_k$ in which each k-by-k submatrix is **TP** (respectively, **TN**).

2.5 Positive Definite Matrices (PD)

Perhaps the most prominent positivity class is defined by the quadratic form for Hermitian matrices. Matrix $A \in M_n(\mathbb{R})$ is called **positive definite** ($A \in$ **PD**) if A is Hermitian ($A^* = A$) and, for all $0 \neq x \in \mathbb{C}^n$, $x^T Ax > 0$.

There are several good sources on **PD** matrices, including [HJ13, Joh70, Bha07].

The **PD** matrices are closed under addition and positive scalar multiplication (they form a cone in $M_n(\mathbb{C})$), and under the Hadamard product, but not under conventional matrix multiplication. They are closed under extraction of principal submatrices and Schur complements. Among Hermitian matrices, all positive eigenvalues, positive leading principal minors, and all principal minors positive are each equivalent to being **PD**. A matrix $A \in M_n(\mathbb{C})$, is **PD** if and only if $A = B^*B$, with $B \in M_n(\mathbb{C})$, nonsingular; B may always be taken to be upper triangular (**Cholesky factorization**).

The standard variation on **PD** is **positive semidefinite** matrices, **PSD**, in which the quadratic form is only required to be nonnegative. **PSD** is the closure of **PD**, and most of the weakened properties follow from this. Of course, **negative definite** (**ND** $=$ $-$**PD**) and **negative semidefinite** (**NSD** $=$ $-$**PSD**) are not generalizations of positivity.

Another important variation is, of course, that the Hermitian part $H(A) = \frac{A+A^*}{2}$ is **PD**. In the case of $M_n(\mathbb{R})$ this just means that the quadratic form is positive, but the matrix is not required to be symmetric. Another variation is that $H(A)$ be **PSD**. There are some references that consider the former class, e.g., [Joh70, Joh72, Joh73, Joh75a, Joh75b, Joh75c, BaJoh76].

2.6 Strictly Copositive Matrices (SC)

An important generalization of **PSD** matrices is the copositive matrices (**C**) for which it is only required that the quadratic form be nonnegative on nonnegative vectors: $A \in M_n(\mathbb{R})$ is **copositive** if $A^T = A$ and $x^TAx \geq 0$ for all $x \geq 0$.

Much theory has been developed about the subtle class of copositive matrices. Because there is no comprehensive reference, Chapter 6 is devoted to this class. Variations include the **strictly copositive** matrices (**SC**) for which positivity of the quadratic form is required on nonnegative, nonzero vectors, and **copositive** $+$, the copositive matrices for which $x \geq 0$ and $x^TAx = 0$ imply $Ax = 0$. Also, the real matrices with Hermitian part in **C** or **SC** could be considered.

2.7 Doubly Nonnegative Matrices (DN)

The intersection of the cone of (symmetric) nonnegative matrices and the cone of **PD** matrices in $M_n(\mathbb{R})$ is the cone of **doubly nonnegative matrices** (**DN**). Natural variations include the closure of **DN** (nonnegative and **PSD**) and the **doubly positive matrices** (**DP**), which are positive and **PD**. The most natural of the three classes, **DN**, lies properly between the other two. Membership in **DN** is straightforward to check, but the generators of the cone are difficult to determine.

2.8 Completely Positive Matrices (CP)

A strong variation on positive semidefinite, as well as (doubly) nonnegative matrices, is the notion of complete positivity: A **completely positive matrix** (**CP**) $A \in M_n(\mathbb{R})$ is a symmetric matrix that can be written as $A = BB^T$ with $B \geq 0$. The minimum number of columns of B in such a representation is the so-called **CP-rank** of A; it may be greater than the rank of A. The **CP** notion strengthens one of the characterizations of **PSD**, namely, that a **PSD** matrix may be written as BB^T (with no other restrictions on B). The **CP** matrices have some clear structure; they form a cone in $M_n(\mathbb{R})$, whose generators are the rank 1, nonnegative **PSD** matrices. However, verifying membership in **CP** is still quite difficult.

There is a book reference [BS-M03] on the **CP** matrices, as well as more recent literature. The **CP** matrices constitute the cone theoretic dual of the copositive matrices, one contained in and the other containing the **PSD** cone.

2.9 P-Matrices (P)

Matrix $A \in M_n(\mathbb{C})$ is a **P-matrix** ($A \in$ **P**), if all its principal minors are positive. Several prior classes, namely, the M-matrices, inverse M-matrices, totally positive, and positive definite matrices are contained in **P**. P-matrices are very important for the linear complementarity problem (and of interest for a variety of other reasons). Despite the fact that the definition involves many fewer minors, checking whether a matrix is a P-matrix, though finite, is generally much more costly than checking whether a matrix is **TP**. Nonetheless, there are important characterizations of P-matrices, covered in Chapter 4. References about properties of P-matrices include [FP62], [FP66], [FP67], and [HJ91].

Classical variations include the $\mathbf{P}_0$-matrices (all principal minors nonnegative) and $\mathbf{P}_k$-matrices, $0 < k \leq n$, in which all k-by-k principal submatrices are P-matrices. Lying strictly between the **TN** matrices and the P-matrices are the intersection of nonnegative matrices and the P-matrices, the nonnegative P-matrices. The positive P-matrices lie between the **TP** matrices and the P-matricces. Thc inverses of M-matrices are contained in the nonnegative P-matrices.

2.10 $\mathbf{H^+}$-Matrices ($\mathbf{H^+}$)

Matrix $A = [a_{ij}] \in M_n(\mathbb{C})$ is **column diagonally dominant** if $|a_{jj}| > \sum_{i=1,\, i\neq j}^{n} |a_{ij}|$ for each $j = 1, 2, \ldots, n$. A is **row diagonally dominant** if A^T is column diagonally dominant. If there is a diagonal matrix D such that AD is row diagonally dominant, then A is called an **H-matrix**, which is equivalent to the existence of a diagonal matrix E such that EA is column diagonally dominant, or that there are diagonal matrices E and D such that EAD is both row and column diagonally dominant. Matrix $A \in M_n(\mathbb{R})$ is an $\mathbf{H^+}$**-matrix** if A is an H-matrix with positive diagonal entries. The M-matrices are contained in the H^+-matrices, and the H^+-matrices are positive stable. The H^+-matrices are of greatest interest when the entries are real.

2.11 Monotone Matrices (Mon)

Matrix $A \in M_{m,n}(\mathbb{R})$ is called **monotone** if whenever $Ax \geq 0$, we have $x \geq 0$. If A is monotone and $Ax = 0$, we also have $A(-x) = 0$, so that $x \geq 0$ and $-x \geq 0$, which means $x = 0$. Thus, A must have linearly independent columns, and if $m = n$, A must be invertible. Thus if $A \in M_n(\mathbb{R})$ is monotone, A^{-1} exists and A^{-1} is nonnegative ($Ax = y \geq 0$). So $Av \geq Au$ implies $v \geq u$, a motivation for the name monotone.

2.12 Semipositive Matrices (SP)

Matrix $A \in M_{m,n}(\mathbb{R})$ is **semipositive** ($A \in$ **SP**) if there is an $x \geq 0$ such that $Ax > 0$
(a weakening of the characterization that characterizes entry-wise positive matrices). Chapter 3 is devoted to their study, and they are the most general class listed.

2.13 Relationships among the Classes

If we interpret each of the above classes in the strict sense over $\mathbb{R}$ and all matrices are square (for example, **DN** means **PD**$\cap$**EP** and $\mathbf{H^+} \subseteq M_n(\mathbb{R})$), then we may give a complete list of containments among the classes. Most of these are straightforward, except that **SP** contains all of the classes, using Theorem 1.4.1 appropriately. Containments are described in the following schema.

$$\begin{array}{llll}
\mathbf{SP} \supseteq \mathbf{P} & \supseteq \mathbf{PD} \supseteq \mathbf{DN} \supseteq \mathbf{CP} \\
& \supseteq \mathbf{TP} \\
& \supseteq \mathbf{H}^{+} \supseteq \mathbf{M} \\
\quad\supseteq \mathbf{SC} & \supseteq \mathbf{PD} \\
\quad\supseteq \mathbf{EP} & \supseteq \mathbf{TP} \\
& \supseteq \mathbf{DN} \\
\quad\supseteq \mathbf{Mon} & \supseteq \mathbf{M}
\end{array}$$

That **SP** $\supseteq$ **P** is proven in Chapter 4, using Theorem 1.4.1 and the characterization that says a real P-matrix is one that cannot weakly reverse all the signs of any real vector. That **SP** $\supseteq$ **SC** uses Theorem 1.4.1 similarly. That **SP** $\supseteq$ **EP** is trivial, as is **SP** $\supseteq$ **Mon**. Other containments are known, using traditional characterizations; all are strict, and there are no other containments.

3

Semipositive Matrices

Our purpose here is to develop the theory of semipositive matrices in some depth. No other comprehensive general reference is available. As seen in Chapter 2, they include the strong versions of all the traditional positivity classes.

3.1 Definitions and Elementary Properties

Recall that $A \in M_{m,n}(\mathbb{R})$ is positive, $A > 0$, if and only if $Ax > 0$ for *every* nonzero $x \geq 0$, $x \in \mathbb{R}^n$. If the quantification on x is changed to *existence*, we have the notion of a semipositive (**SP**) matrix.

Definition 3.1.1 For $A \in M_{m,n}(\mathbb{R})$, let

$$S_+(A) = \{x \in \mathbb{R}^n \colon x \geq 0 \text{ and } Ax > 0\} \quad \text{and} \quad R_+(A) = A\,S_+(A).$$

Of course, $R_+(A) \subseteq \mathbb{R}^m_+$ (and its interior), $S_+(A) \subseteq \mathbb{R}^n_+$, the nonnegative orthants in $\mathbb{R}^m$ and $\mathbb{R}^n$, respectively. Also $R_+(A) \neq \emptyset$ if and only if $S_+(A) \neq \emptyset$. If $S_+(A) \neq \emptyset$, then A is called **semipositive** ($A \in$ **SP**).

Note that, by continuity, if A is **SP**, then there exists $x > 0$ such that $Ax > 0$.

Definition 3.1.2 In a similar way, we may define

- **seminonnegative** (**SNN**) if there exists nonzero $x \geq 0$ such that $Ax \geq 0$;
- **seminegative** (**SN**): $-A$ is **SP**;
- **seminonpositive** (**SNP**): $-A$ is **SNN**;
- **semizero** (**SZ**): A has a nonnegative, nonzero null vector.

We note that

Remark 3.1.3 If A is **SP**, then $S_+(A)$ is a nonempty cone in $\mathbb{R}^n_+$ and $R_+(A)$ is a nonempty cone in $\mathbb{R}^m_+$.

Of course, there are also *left* versions of these concepts, in which vectors are multiplied on the left. For example, **left semipositive** (left **SP**, **LSP**) means that there is a $y \geq 0$ such that $y^T A > 0$, and a prefix "L" means the left-hand version; equivalently A^T is **SP**.

Definition 3.1.4 We call a matrix $C \in M_{r,m}(\mathbb{R})$ **row positive** if $C \geq 0$ has at least one positive entry in each row. Note that the positive matrices are exactly those that map the positive orthant of $\mathbb{R}^m$ to that of $\mathbb{R}^r$. Recall that $B \in M_{n,s}(\mathbb{R})$ is **monotone** if for $x \in \mathbb{R}^s$, $Bx \geq 0$ implies $x \geq 0$. If $s = n$, then B is invertible and $B^{-1} \geq 0$ (see Section 2.11). Note also that if $A \geq 0$ is invertible, then both A and A^T are row positive.

Theorem 3.1.5 If $A \in M_{m,n}(\mathbb{R})$ is **SP**, then CAB is **SP** whenever $C \in M_{r,m}(\mathbb{R})$ is row positive and $B \in M_n(\mathbb{R})$ is monotone.

Proof If A is **SP**, let $x \in S_+(A)$. Choose $u = B^{-1}x \in \mathbb{R}^q$ so that $Bu = x$; then $u \geq 0$. Now $CABu = CAx = Cy$ with $y > 0$. Because C is row positive, $Cy > 0$, and, thus, $u \in S_+(CAB)$, which means CAB is **SP**. □

Observation 3.1.6 If $A \in M_{m,n}(\mathbb{R})$ and $A \geq 0$, then A is **SP** if and only if A is row positive. Moreover, if A is row positive, $S_+(A)$ includes the entire positive orthant of $\mathbb{R}^n$. Also, if $A \in M_n(\mathbb{R})$ is monotone, then A is **SP**.

Furthermore, no broader classes than those mentioned preserve **SP**. This allows us to identify the most natural automorphisms of **SP** matrices: permutation equivalence and positive diagonal equivalence. Let P_n denote the n-by-n permutation matrices, D_n the invertible n-by-n diagonal matrices and D_n^+ the positive diagonal matrices in D_n. Note that P_n and D_n^+ are both contained in the row positive and monotone matrices; so are the **monomial matrices**, $P_n D_n^+$, which are the n-by-n matrices with exactly one positive entry in each row and column.

Corollary 3.1.7 Let $A \in M_{m,n}(\mathbb{R})$ and let $P \in P_m$ and $Q \in P_n$. Then, PAQ is **SP** if and only if A is **SP**.

Proof Because $A = P^T(PAQ)Q^T$, it suffices to show that A being **SP** implies that PAQ is **SP**. But, because P is row positive and Q is monotone, this follows from Theorem 3.1.5. □

Corollary 3.1.8 Let $A \in M_{m,n}(\mathbb{R})$, and let $D \in D_m^+$ and $E \in D_n^+$. Then DAE is **SP** if and only if A is **SP**.

Proof Because D_n^+ is inverse closed and $A = D^{-1}(DAE)E^{-1}$, it suffices to show that A being **SP** implies that DAE is **SP**. But, again, as D is row positive and E is monotone, this follows from Theorem 3.1.5. □

Remark 3.1.9 If $A \in M_{m,n}(\mathbb{R})$ and R and S are monomial matrices of appropriate size, then RAS is **SP** if and only if A is **SP**. This follows from the preceding two facts.

Some simple consequences of the definition, through $S_+(A)$, include the following.

Let $E_{ij} \in M_n(\mathbb{R})$ denote a matrix whose (i,j) entry equals 1 and all other entries are zero. If $A \in M_{m,n}(\mathbb{R})$ is **SP** and $B_t = A + tE_{ij}$, then $S_+(B_{t_1}) \subseteq S_+(B_{t_2})$ for $t_1 \leq t_2$, so that B_t is **SP** for all $t \geq 0$. In fact, for $t > 0$, $S_+(A) \subseteq S_+(B_t)$, with strict containment, unless $A > 0$ and the interior of $S_+(A)$ is already the positive orthant. Also, if A is **SP**, then each row of A must have at least one positive entry, else $S_+(A) = \emptyset$.

Rows and columns play different and essentially dual roles in **SP** matrices. Let us refer to a submatrix of $A \in M_{m,n}(\mathbb{R})$ obtained by deleting a column (row) as a **column- (row-) deleted submatrix** of A.

Theorem 3.1.10 Let $A \in M_{m,n}(\mathbb{R})$.

(1) If there is a column-deleted submatrix of A that is **SP**, then A is **SP**.
(2) If A is **SP**, then any nonempty, row-deleted submatrix of A is **SP**.

Proof
(1) If B is a column-deleted submatrix of A, we may suppose, by Corollary 3.1.7, that B occupies the first k columns of A. Then, if $x \in S_+(B) \subseteq \mathbb{R}^k$, and $[x \quad 0]^T \in S_+(A)$, which is nonzero, as x is. The augmentation is with $n-k$ zeroes.

(2) If B is a row-deleted submatrix of A, then $S_+(A) \subseteq S_+(B)$, so that $S_+(B)$ is nonempty and B is **SP**. □

Remark 3.1.11 Because $A \in M_{m,1}(\mathbb{R})$ with positive entries (a positive column) is **SP**, it follows from Theorem 3.1.10 (1) that any matrix with a positive column is **SP**.

Definition 3.1.12 If A is **SP**, but no proper, column-deleted submatrix of A is **SP**, then we call A **minimally SP (MSP)**. This can happen only when $m \geq n$, and we shall study the notion in greater depth in Section 3.5. Corresponding to Theorem 3.1.10 (2), rows may always be added to an **SP** matrix to produce a "taller" one. Of course, a row-deleted submatrix of A may be **SP**, even when A is not **SP**.

We may now give a long list of conditions equivalent to **SP** (and there are analogous equivalent statements for the other concepts).

Theorem 3.1.13 The following statements about $A \in M_{m,n}(\mathbb{R})$ are equivalent:

(1) A is **SP**;
(2) the cone generated by the columns of A intersects the positive orthant in $\mathbb{R}^m$;
(3) there is an $x \in \mathbb{R}^n$ such that $x > 0$ and $Ax > 0$;
(4) $S_+(A)$ has nonempty interior;
(5) there is a $b \in \mathbb{R}^m$, $b > 0$, such that $Ax = b$ has a positive solution;
(6) there is a positive diagonal matrix D such that AD has positive row sums;
(7) there exist positive diagonal matrices D and E such that EAD has constant positive row sums;
(8) A^T is not **SNP**.

Proof Any element of $S_+(A)$ gives a linear combination of the columns of A that is in the positive orthant, so (1) implies (2). A sufficiently small positive perturbation of the coefficients of the linear combination shows that (2) implies (3). An x satisfying (3) lies in the interior of $S_+(A)$, so that (3) implies (4). Choose b in the image of the interior of $S_+(A)$ to see that (4) implies (5). Let $D = \operatorname{diag}(x)^{-1}$ to show that (5) implies (6). If b is the row sum vectors of AD, let $E = \operatorname{diag}(b)^{-1}$ to show that (6) implies (7). Because of (7) the first alternative of Theorem 1.4.1 is satisfied. The fact that the second alternative cannot be satisfied is statement (8). That (8) implies (1) is another application of the Theorem of the Alternative, to complete the proof. □

That $S_+(A)$ is nonempty means that it has a nonempty interior. A corresponding statement is not so for **SNN** or **SNP** matrices.

Example 3.1.14 Let $A = \begin{bmatrix} 1 & -1 \\ -1 & 1 \end{bmatrix}$. Then A is **SNN** (use $x = [1 \quad 1]^T$), but not **SP**. If $x = [x_1 \quad x_2]^T$, then $Ax = \begin{bmatrix} x_1 - x_2 \\ x_2 - x_1 \end{bmatrix}$. If $x \geq 0, \neq 0$, then at least one component of Ax is negative, unless $x_1 = x_2$, in which case $Ax = 0$. The set $\{x\colon x_1 = x_2 \geq 0\}$ has an empty interior in $\mathbb{R}^2$.

Example 3.1.15 Let $A = \begin{bmatrix} 1 & -1 \\ 1 & -1 \end{bmatrix}$. Then A is **SP**, as $\begin{bmatrix} 1 \\ 0 \end{bmatrix} \in S_+(A)$, but A^T is not **SP**. So it is possible to be **SP** and not left **SP**, or vice versa.

Definition 3.1.16 If A is both **SP** and left **SP** (**LSP**), then it is called **symmetrically SP**. A symmetric matrix that is **SP** is, of course, symmetrically **SP**, but a symmetrically **SP** matrix need not be symmetric.

We note that if $A \in M_n(\mathbb{R})$ is **SP** and invertible, then A^{-1} is also **SP**.

Theorem 3.1.17 Suppose that $A \in M_n(\mathbb{R})$ is invertible. Then A is **SP** if and only if A^{-1} is **SP**.

Proof Because of Theorem 3.1.13 (3), an **SP** matrix A is one such that there are $x, y > 0$ with $y = Ax$. But then $x = A^{-1}y$ and vice versa. □

Of course, a square **SP** matrix need not be invertible, e.g., $\begin{bmatrix} 1 & 1 \\ 1 & 1 \end{bmatrix}$.

3.2 Sums and Products of SP Matrices

If $A,B \in M_{m,n}(\mathbb{R})$ are **SP**, then $A + B$ is defined, and we may ask (1) if it is **SP** and (2) which matrices in $M_{m,n}(\mathbb{R})$ may be generated in this way? If $A \in M_{m,n}(\mathbb{R})$ and $B \in M_{n,p}(\mathbb{R})$ are **SP**, then AB is defined, and we may ask (1) if AB is **SP** and (2) which matrices in $M_{m,p}(\mathbb{R})$ may be generated in this way? Not surprisingly, neither the sum, nor the product, of two **SP** matrices need to be **SP**.

Example 3.2.1 Let

$$A = \begin{bmatrix} 1 & -5 \\ 0 & 1 \end{bmatrix}, \quad B = \begin{bmatrix} 1 & 1 \\ 1 & 1 \end{bmatrix}.$$

Then A, A^T and B are all **SP**, as $A\begin{bmatrix} 6 \\ 1 \end{bmatrix} = \begin{bmatrix} 1 \\ 1 \end{bmatrix}$, $A^T\begin{bmatrix} 6 \\ 1 \end{bmatrix} = \begin{bmatrix} 1 \\ 1 \end{bmatrix}$ and $B > 0$. However,

$$A + A^T = \begin{bmatrix} 2 & -5 \\ -5 & 2 \end{bmatrix}$$

is not **SP**, as no nonnegative combination of its columns can be positive, and

$$AB = \begin{bmatrix} -4 & -4 \\ 1 & 1 \end{bmatrix}$$

is not **SP**, as it has a negative row.

In order to give sufficient conditions so that the product or sum of **SP** matrices be **SP**, we exploit $S_+(A)$ and $R_+(A)$.

Now, we may give a general sufficient condition for a product of **SP** matrices to be **SP**.

Theorem 3.2.2 Let $A \in M_{m,n}(\mathbb{R})$, $C \in M_{r,m}(\mathbb{R})$, and $B \in M_{n,s}(\mathbb{R})$ be **SP**. Then,

(1) if $R_+(B) \cap S_+(A) \neq \emptyset$, then AB is **SP**;
(2) if $A\,(R_+(B) \cap S_+(A) \cap R_+(C)) \neq \emptyset$, then CAB is **SP**.

Proof
(1) If $u \in R_+(B) \cap S_+(A)$, then $u = Bx > 0$ for some $x \geq 0$, and $Au > 0$. Thus $ABx = Au > 0$, and so AB is **SP**.

(2) Let $z \in A(R_+(B) \cap S_+(A) \cap R_+(C))$. Then $z = Au$, where $u \in R_+(B) \cap S_+(A) \cap R_+(C)$. Because $u \in R_+(B) \cap S_+(A)$, as in (1), we have $u = Bx > 0$ for some $x \geq 0$. It follows that $CABx = CAu = Cz > 0$ and so CAB is **SP**. □

Of course, it may happen that (a) one or both of A and B are not **SP**, or (b) that both are, and the sufficient conditions of Theorem 3.2.2 are not met, and yet AB still be **SP**.

Example 3.2.3 Consider

$$A = \begin{bmatrix} -1 & -1 \\ -1 & 0 \end{bmatrix}, \ B = \begin{bmatrix} -1 & -1 \\ 0 & 1 \end{bmatrix}, \ AB = \begin{bmatrix} 1 & 0 \\ 1 & 1 \end{bmatrix}.$$

Note that as A and B are not **SP**, $R_+(B) \cap S_+(A) = \emptyset$, yet $AB \in$ **SP**. Also, for

$$C = \begin{bmatrix} 1 & 1 \\ 0 & 1 \end{bmatrix}, \ E = \begin{bmatrix} 1 & 0 \\ 1 & 0 \end{bmatrix}, \ CE = \begin{bmatrix} 2 & 0 \\ 1 & 0 \end{bmatrix}.$$

Note that $S_+(C)$ comprises the nonnegative vectors with positive second entry, $S_+(E)$ comprises the nonnegative vectors with positive first entry, and thus $R_+(E) = ES_+(E)$ comprises the vectors with positive first entry and zero second entry. Thus $R_+(E) \cap S_+(C) = \emptyset$, yet $CE \in$ **SP**.

It seems not so straightforward to characterize, when AB is **SP** in terms of individual characteristics of A and B. The interaction of characteristics is key.

If we begin with a general matrix in $M_{m,n}(\mathbb{R})$ and ask if it is the product of two **SP** matrices, the answer is clearer, though nontrivial. To see what happens, a technical lemma is useful.

Lemma 3.2.4 For $m \geq 2$, $n \geq 1$, suppose that $C \in M_{m,n}(\mathbb{R})$. If $\{v, w\} \subseteq \mathbb{R}^m$ is a linearly independent set and $v_2 \in R^n$ is such that $\{Cv_2, w\}$ is a linearly independent set, then there exist $A \in M_m(\mathbb{R})$ and $B \in M_{m,n}(\mathbb{R})$ such that $Av_1 = w$, $Bv_2 = w$ and $C = AB$.

Proof Choose $A \in M_n(\mathbb{R})$ to be an invertible matrix such that $Av_1 = w$ and $Aw = Cv_2 \neq 0$. Set $B = A^{-1}C$, so that $C = AB$. Then, $Bv_2 = A^{-1}Cv_2 = w$ and the stated requirements are fulfilled. □

Theorem 3.2.5 If $m \geq 2$, $n \geq 1$, and $C \in M_{m,n}(\mathbb{R})$, then there exist $A \in M_m(\mathbb{R})$ and $B \in M_{m,n}(\mathbb{R})$, both **SP**, such that $C = AB$.

Proof If rank $C \geq 1$, positive vectors v_1, v_2, and w may be chosen, so as to fulfill the hypothesis of Lemma 3.2.4. The positivity of these vectors means that

A and B are **SP**, completing the proof of this theorem. If $C = 0$, and $m \geq 2$, we may choose

$$A = \begin{bmatrix} 1 & -1 & 0 & \cdots & 0 \\ 1 & -1 & 0 & \cdots & 0 \\ \vdots & \vdots & \vdots & \vdots & \vdots \\ 1 & -1 & 0 & \cdots & 0 \end{bmatrix}, \quad B = \begin{bmatrix} 1 & 1 & 1 & \cdots & 1 \\ 1 & 1 & 1 & \cdots & 1 \\ \vdots & \vdots & \vdots & \vdots & \vdots \\ 1 & 1 & 1 & \cdots & 1 \end{bmatrix}$$

of the required sizes, so that $C = AB$. Because A and B both have positive columns, they are **SP**. □

Remark 3.2.6 The requirement that $m \geq 2$ in Theorem 3.2.5 is necessary. Suppose that $C \in M_{1,n}(\mathbb{R})$ and all the entries of C are negative. If $C = AB$ and A is 1-by-1, then either the lone entry of A is negative, in which case A is not **SP**, or B has all negative entries and is therefore not **SP**. However, if $C \in M_{1,n}(\mathbb{R})$, $n \geq 2$, either C has some positive entries and is, itself, **SP**, in which case $C = CI$ is an **SP** factorization of C, or $C \leq 0$. If $C \leq 0$, then we may choose $A = [1 \ \ -1]$ and $B > 0$. $B \in M_{2,n}(\mathbb{R})$ so that $AB = C$ with both A and B **SP**. Of course, if $m, n = 1$ and $C \leq 0$, then $C = AB$, with both $A, B \in M_{1,1}(\mathbb{R})$ and **SP**, is not possible. But $C = AB$, with $A \in M_{1,2}(\mathbb{R})$ and $B \in M_{2,1}(\mathbb{R})$ and both A, B **SP** is possible. Suppose $C = (-c)$ with $C > 0$. Then let $A = [1 \ \ -1]$ and $B = \begin{bmatrix} 1 \\ 1-c \end{bmatrix}$.

Together with Theorem 3.2.5, the latter part of the remark above gives the following complete result.

Theorem 3.2.7 Suppose that $C \in M_{m,n}(\mathbb{R})$. Then, there is a $k \geq 1$, and $A \in M_{m,k}(\mathbb{R})$ and a $B \in M_{k,n}(\mathbb{R})$, both **SP** such that $C = AB$. If $m = 1$ and C has no positive entries, we must have $k \geq 2$ and $k = 2$ suffices. Otherwise, $k \leq m$ is possible.

For the sum of two **SP** matrices there are similar sufficient conditions.

Lemma 3.2.8 If $A, B \in M_{m,n}(\mathbb{R})$ are **SP**, then $S_+(A) \cap S_+(B) \subseteq S_+(A + B)$.

Proof Let A, B be **SP** and $x \in S_+(A) \cap S_+(B)$. Then $(A + B)x = Ax + Bx > 0$ and so $x \in S_+(A + B)$. □

Theorem 3.2.9 If A,$B \in M_{m,n}(\mathbb{R})$ are **SP**, and if $S_+(A) \cap S_+(B) \neq \emptyset$, then $A + B$ is **SP**.

Proof This follows directly from the above lemma. □

Of course, again this sufficient condition is not necessary.

Example 3.2.10 Consider

$$A = \begin{bmatrix} -1 & -1 \\ 2 & 2 \end{bmatrix},\ B = \begin{bmatrix} 2 & 2 \\ -1 & -1 \end{bmatrix},\ A + B = \begin{bmatrix} 1 & 1 \\ 1 & 1 \end{bmatrix}.$$

Note that as A and B are not **SP**, $S_+(A) \cap S_+(B) = \emptyset$, yet $A + B \in$ **SP**.

Now, which matrices occur as the sum of two **SP** matrices? In this case, there is no flexibility in the dimensions of the summands; they must be exactly the same as the desired sum. This means that there are two restrictions.

Theorem 3.2.11 Suppose that $C \in M_{m,n}(\mathbb{R})$, with $n \geq 2$. Then, there exist **SP** matrices $A, B \in M_{m,n}(\mathbb{R})$ such that $C = A + B$.

Proof First, pick A so that the entries of the first column are all 1s, the second column is the second column of C with one subtracted from each entry, and all of the other columns match the columns of C. Then set the first column of B to be the first column of C with 1 subtracted from each entry, let the second column have every entry equal to 1, and make all of the other columns have entries equal to 0. Then $C = A + B$ and A and C are both SP, as each has a positive column. □

Remark 3.2.12 The requirement that $n \geq 2$ in Theorem 3.2.11 is necessary. If $n = 1$ and C has a negative entry, then for $C = A + B$, A or B must have a negative entry in the same position, thus a negative row, and not be **SP**. So, when $n = 1$, C is the sum of two **SP** matrices if and only if A itself is **SP**.

Despite it being easy to understand when a matrix is the sum of two **SP** matrices, it cannot so often be done if both left and right **SP** be required of the summands. For sums, the 0 matrix of any size is an easy counterexample. If $A+B = 0$ and A and B are both left **SP** and **SP**, then $A = -B$, so that A is both left and right **SN** as well. However, then A is **SP** and left **SN**, and therefore left **SNP** as well, which contradicts the Theorem of the Alternative 1.4.1.

3.3 Further Structure and Sign Patterns of SP Matrices

If $A \in M_{m,n}(\mathbb{R})$ and $B \in M_{p,n}(\mathbb{R})$, we may form

$$A \oslash B = \begin{bmatrix} A \\ B \end{bmatrix} \in M_{m+p,n}(\mathbb{R}).$$

Because of Theorem 3.1.10, in order for $A \oslash B$ to be **SP**, we must have both A and B be **SP**. If A and B are both **SP**, what further do we need in order that $A \oslash B$ be **SP**?

Lemma 3.3.1 For $A \in M_{m,n}(\mathbb{R})$ and $B \in M_{p,n}(\mathbb{R})$,

$$S_+(A \oslash B) = S_+(A) \cap S_+(B).$$

Proof Observe that $x \in S_+(A \oslash B)$ if and only if $x \geq 0$, $Ax > 0$ and $Bx > 0$. □

From this follows a useful observation.

Theorem 3.3.2 If $A \in M_{m,n}(\mathbb{R})$ and $B \in M_{p,n}(\mathbb{R})$ are **SP**, then $A \oslash B$ is **SP** if and only if $S_+(A) \cap S_+(B) \neq \emptyset$.

Proof By definition, $A \oslash B$ is **SP** when $S_+(A \oslash B) \neq \emptyset$; the result follows from the above lemma. □

If $p = m$, we note that the condition in Theorem 3.3.2 is the same as the sufficient condition for $A + B$ to be **SP** in Theorem 3.2.9. This means that

Corollary 3.3.3 If $A, B \in M_{m,n}(\mathbb{R})$ are **SP**, and $A \oslash B$ in **SP**, then $A + B$ is **SP**.

This could also be proven by writing $A + B$ as $[I\ I](A \oslash B)$ and applying Theorem 3.2.2 (1). Thus, products, sums, and the operator $\oslash$ are closely related. Another view would be the intersection of half-spaces; see Section 1.3.3.

Of course, Theorem 3.3.2 shows how the semipositive cone is diminished as we add rows. If we add positive rows, there is no diminution at all, as the semipositive cone applied to a positive row is the entire nonnegative orthant, save the 0 vector. This means

Corollary 3.3.4 Suppose that $A \in M_{m,n}(\mathbb{R})$ and $B \in M_{p,n}(\mathbb{R})$ with $B > 0$. Then $A \oslash B$ is **SP** if and only if A is **SP**.

What, then, about matrices of the form $[A\ B]$? Suppose that $A \in M_{m,n}(\mathbb{R})$ and $B \in M_{m,q}(\mathbb{R})$, and we consider $A \ominus B = [A\ B]$. An important fact is the following.

Theorem 3.3.5 Suppose $A \in M_{m,n}(\mathbb{R})$ and $B \in M_{m,q}(\mathbb{R})$. If A or B is **SP**, then $A \ominus B$ is **SP**.

Proof If A is **SP**, let $x \in S_+(A)$ and define $y = \left[\begin{smallmatrix} x \\ 0 \end{smallmatrix}\right] \in \mathbb{R}^{n+q}$. Then $y \geq 0$ and $(A \ominus B)y = Ax > 0$, and thus $A \ominus B$ is **SP**. The proof when B is **SP** is similar. □

By Corollary 3.1.7, if any subset of the columns of $A \in M_{m,n}(\mathbb{R})$ forms an **SP** matrix, then it follows from Theorem 3.3.5 that A is **SP**. This has already been mentioned in Theorem 3.1.10 (1). Because a positive column vector is trivially an **SP** matrix, a special case is mentioned in Remark 3.1.11.

Corollary 3.3.6 If $A \in M_{m,n}(\mathbb{R})$ has a positive column, then A is **SP**.

The previous corollary has an interesting and far-reaching generalization. We say that $A \in M_{m,n}(\mathbb{R})$ has a **positive front** if the columns may be permuted so that, in each row, there is a nonzero entry and the first nonzero entry is positive. These positive entries are called **fronted positives**. Of course, a matrix with a positive column has a positive front, but it may happen that there is a positive front without a positive column.

Example 3.3.7 Let

$$A = \begin{bmatrix} 1 & 0 & -1 \\ 0 & -1 & 1 \\ 0 & -1 & 1 \end{bmatrix}.$$

Then the interchange of columns 2 and 3 yields

$$A' = \begin{bmatrix} 1 & -1 & 0 \\ 0 & 1 & -1 \\ 0 & 1 & -1 \end{bmatrix},$$

so that A has a positive front. It is easy to see that A' is **SP** as

$$\begin{bmatrix} 2 \\ 1 \\ 0 \end{bmatrix} \in S_+(A')$$

and so A is **SP** by Corollary 3.1.7.

It turns out that the existence of a positive front, like a positive column, is always sufficient for **SP**. As we shall see, this is a minimal condition in terms of the sign pattern alone.

Theorem 3.3.8 If $A \in M_{m,n}(\mathbb{R})$ has a positive front, then A is **SP**.

Proof If A can be permuted to a matrix A' that has a positive front, by Corollary 3.1.7, it is enough to show that A' is semipositive because semipositivity is not changed under permutation. If A' has z zero columns as its first z columns, set the first z entries of v to be zero. Set the remaining d entries of v as $x^d, x^{d-1}, \dots, x$, where x is a fixed number to be specified later. Each row of the product $A'v$ will be a polynomial of degree less than or equal to d, whose first entry is positive (because the first nonzero entry in each row is positive, and there are no zero rows because A has a positive front). Because the coefficient of the monomial with the largest degree in each of these polynomials is positive, the limit as x approaches infinity for all of these polynomials is infinity. Therefore there is some $x > 0$ where each of these polynomials are greater than zero, which implies that A' is semipositive, and thus A is semipositive. □

We note that it is not difficult to test for the existence of a positive front [DGJT16]. An algorithm that does this is described next.

Algorithm PF

To determine if an m-by-n matrix A has a positive front, first check that the matrix has no zero row. If it does, this matrix does not have a positive front. If it does not, construct a sequence of matrices with $A_1 := A$. Construct A_2 by removing the 0 columns in A_1. Afterwards for $k > 1$, construct A_{k+1} recursively in the following fashion:

1. Determine if A_k contains a column without a negative entry. If all p columns of A_k contain a negative entry, this matrix does not have a positive front. Otherwise, choose a j-th column of A_k, which we will denote by a_{kj}, which has no negative entries. Set A_{k+1} by deleting the j-th column of A_k, as well as any row in A_k in which a_{kj} contains a positive entry (leaving the rest of the matrix), and repeat step 1, unless this deletion will result in the loss of the entire matrix. If this is the case, set A_k as the "final matrix" and proceed to step 2.

2. Define a function $C(a_{kj})$ that takes the vector a_{kj} and returns the original column of A associated with a_{kj}.

3. If A has r columns of entirely 0, set the first r columns of A' to be 0. Then, set the next column to be $C(a_{3j})$, the next column to be $C(a_{4j})$, ..., and the next column to be $C(a_{kj})$, where A_k is the final matrix. If there are columns of A that have not been mapped to by this C function and are nonzero, set them in an arbitrary order after the $C(a_{kj})$ column. Note that this process must terminate in at most n steps because after n repetitions of step 1, the entire matrix will be "deleted."

Algorithm PF terminates (in A') if and only if A has a positive front [DGJT16, theorem 4.5].

By the **sign pattern** of $A = [a_{ij}] \in M_{m,n}(\mathbb{R})$, we mean an m-by-n array $\mathcal{A}$ of signs $(+, -, 0)$ in which $a_{ij} > 0\,(<, = 0)$ if and only if the (i,j) entry of a is $+(-, 0)$. We may think of the sign pattern $\mathcal{A}$ as both the designator of the set of all matrices with this sign pattern and as a combinational object that may be manipulated like a matrix in many situations. Note that the notion of a positive front makes sense for a sign pattern as well as a matrix. What sign patterns, then, **require SP** (i.e., all matrices of this sign pattern are **SP**)?

Of course, we now know that those with a positive front do. Interestingly, they are the only ones.

Theorem 3.3.9 An m-by-n sign pattern requires **SP** if and only if it has a positive front.

Proof Theorem 3.3.8 gives us sufficiency. We only need to show that, if a sign pattern requires **SP**, then it has a positive front. To show this, assume that $A \in M_{m,n}(\mathbb{R})$ does not have a positive front. The first possibility is that A has a zero row, in which case A^T has a zero column and thus is not seminonpositive (because if A^T has an i-th 0 column, the vector with a 1 in the i-th entry and 0s elsewhere will yield a zero vector), so A is not semipositive. The other possibility is that at some point in Algorithm PF above, at the matrix A_k (a p-by-q matrix) there is a negative entry in each column. Let $C(a_{kj})$ be defined as in step 2 of Algorithm PF above. Construct a row vector v^T which is 0 in the i-th row if the i-th row of A was previously deleted in the algorithm, and 1 otherwise. Choose a matrix B to have the same sign pattern as A, but so that the column sums of any column with a negative entry is zero. Therefore $v^T B = 0$, but v^T is nonzero because if every entry were zero, every row would have been deleted and this algorithm would have terminated. Therefore B is left seminonpositive and thus not semipositive; therefore the sign pattern of A does not require semipositivity. □

We may also ask which sign patterns **allow** **SP**, i.e., there are **SP** matrices in the sign pattern class. Of course, it is necessary that each row contains at least one $+$. But, if there is a $+$ in every row, we may choose a matrix of the sign pattern with a large positive entry in each row of A and all other entries sufficiently small so that $Ae > 0$, a sort-of diagonal dominance. This gives the following theorem.

Theorem 3.3.10 An m-by-n sign pattern allows **SP** if and only if each row contains a $+$.

3.4 Spectral Theory of SP Matrices

If we consider only square matrices $A \in M_n(\mathbb{R})$, it is natural to ask what is special about the spectra of **SP** matrices, either just the multi-set of eigenvalues, or, more precisely, the similarity class (Jordan) structure. Interestingly, there is almost nothing special, which we show here. Nearly all matrices in $M_n(\mathbb{R})$ are similar to an **SP** matrix.

Lemma 3.4.1 Every non-scalar matrix in $M_n(\mathbb{R})$ is similar to a real matrix with positivc first column.

Proof Unless one is scalar and the other is not, two matrices in $M_2(\mathbb{R})$ are similar if and only if they have the same trace and determinant. Furthermore, unless the diagonal entries are the eigenvalues, both off-diagonal entries will be nonzero. Thus, a simple calculation shows that the claim of the lemma is valid in the 2-by-2 case. Now, let $A \in M_n(\mathbb{R})$ be non-scalar. It is clearly similar to a non-diagonal matrix $B \in M_n(\mathbb{R})$, which, in turn, has a 2-by-2 non-scalar principal submatrix that we may assume, without loss of generality, is in the first two rows and columns and has a positive first column.

Now, if the remaining entries in the first column of B are all nonzero, the proof is completed by performing similarity on B by a diagonal matrix of $\pm$ 1s, so as to adjust the signs of the entries in the first column to be all positive and complete the proof. If not all entries in the first column are nonzero, we may make them so, by a sequence of similarities of the form

$$\begin{bmatrix} I & 0 & 0 \\ 0 & \begin{bmatrix} 1 & 0 \\ 1 & 1 \end{bmatrix} & 0 \\ 0 & 0 & I \end{bmatrix},$$

in which the second I may not be present. Then, we may proceed as in the case of a totally nonzero first column to complete the proof. $\square$

This is the key to the observation of interest.

Theorem 3.4.2 Except for nonpositive scalar matrices, every matrix in $M_n(\mathbb{R})$ is similar to an **SP** matrix.

Proof If $A \in M_n(\mathbb{R})$ is not a scalar matrix, then according to Lemma 3.4.1, there is a matrix with positive column in its similarity class. By Corollary 3.3.6, this matrix is **SP**. Because any positive diagonal matrix is **SP**, positive scalar matrices are, as well, which completes the proof. $\square$

From this we have

Corollary 3.4.3 Let n be given and suppose that $\Lambda = \{\lambda_1, \ldots, \lambda_n\}$ is a multi-set of complex numbers that is the spectrum of a real matrix. Then Λ is the spectrum of an **SP** matrix in $M_n(\mathbb{R})$, unless $n = 1$ and $\lambda_1 \leq 0$.

Recall that a multi-set $\Lambda = \{\lambda_1, \ldots, \lambda_n\}$ of complex numbers is the spectrum of a real matrix if the multiplicity in Λ of any non-real complex number is the same as its conjugate.

3.5 Minimally Semipositive Matrices

Recall that $A \in M_{m,n}(\mathbb{R})$ is called **minimally semipositive** (**MSP**) if it is **SP** and no column-deleted submatrix is **SP**. Otherwise, an **SP** matrix is called **redundantly semipositive** (**RSP**). These two concepts may be viewed in terms of $S_+(A)$.

Lemma 3.5.1 If $A \in M_{m,n}(\mathbb{R})$ is **SP**, then A is **RSP** if and only if there are vectors in $S_+(A)$ with 0 components. Column j may be deleted from A to leave an **SP** matrix if and only if there is a vector in $S_+(A)$ whose j-th component is 0. Conversely, A is **MSP** if and only if all components of vectors in $S_+(A)$ are positive.

Proof There is an $x \in S_+(A)$ with $x_j = 0$ if and only if column j of A is redundant to the semipositivity of A. Both claims of the lemma follow from this. □

Example 3.5.2 Matrix $A = \begin{bmatrix} 2 & 2 & -3 \\ -1 & 2 & 0 \\ 0 & -2 & 3 \end{bmatrix}$ is **SP**, as $e = [1 \quad 1 \quad 1]^T \in S_+(A)$. But, deletion of column 2 or 3 leaves a matrix with a row of nonpositive numbers. Deletion of column 1 leaves a matrix with the square submatrix $\begin{bmatrix} 2 & -3 \\ -3 & 2 \end{bmatrix}$. Because the second row is the negative of the first, it cannot be **SP**, so that deletion of column 1 also leaves a matrix that is not **SP**. Thus, A is **MSP**.

Notice that for the matrix A above,

$$A^{-1} = \begin{bmatrix} 1/2 & 0 & 1/2 \\ 1/4 & 1/2 & 1/4 \\ 1/6 & 1/3 & 1/2 \end{bmatrix},$$

so that not only is A invertible, but $A^{-1} \geq 0$.

An **SP** matrix may have any disparity between the number of rows and columns. In trying to understand the special structure of **MSP** matrices, an important fact is that there cannot be more columns than rows. But, first we notice that any **SP** matrix with a nontrivial null space must be **RSP**.

Lemma 3.5.3 Suppose that $A \in M_{m,n}(\mathbb{R})$ is **SP** and that $\mathrm{Null}(A) \neq \{0\}$. Then A is **RSP**.

Proof Let $x \neq 0$ be an element of $\mathrm{Null}(A)$, $y \in S_+(A)$, and define $y(t) = y + tx$. Note that $Ay(t) = Ay > 0$. If $y \not> 0$, we are done, using Lemma 3.5.1. If $y > 0$, then $y(t) \geq 0$ for sufficiently small t, i.e., there is a $T > 0$ such that for $0 \leq t \leq T$, $y(t) \geq 0$. Because we may assume without loss of generality that x has some negative entries (else, replace x with $-x$), there is a $t > 0$ such that $y(t) \geq 0$ and $y(t) \not> 0$. But, then $y(t) \in S_+(A)$ has 0 entries, so that, by Lemma 3.5.1 again, A is **RSP**. □

Remark 3.5.4 Observe that if $A \in M_n(\mathbb{R})$ is invertible and $A^{-1} \geq 0$, then A^{-1} is row positive (no row can be 0, as $\text{rank}(A^{-1}) = n$). Furthermore, if A^{-1} is row positive and $x > 0, x \in \mathbb{R}^n$, then $A^{-1}x > 0$.

It follows from Lemma 3.5.3 that columns may be deleted from an **RSP** matrix so that an **SP** matrix with linearly independent columns results. Such an **SP** matrix may or may not be **MSP**.

Example 3.5.5 Let $A = \begin{bmatrix} 1 & 1 \\ 1 & 2 \\ 1 & 3 \end{bmatrix}$. Then, A is **SP** with linearly independent columns, but A is also **RSP**, as both its columns are positive.

The important consequence of Lemma 3.5.3 is the following.

Theorem 3.5.6 If $A \in M_{m,n}(\mathbb{R})$ is **MSP**, then A has linearly independent columns, so that $n \leq m$.

Proof The linearly independent columns follow from Lemma 3.5.3, and then $n \leq m$ follows as well. □

Another useful fact has a proof similar to that of Lemma 3.5.3, though it is independent.

Lemma 3.5.7 Let $A \in M_{m,n}(\mathbb{R})$ be **SP**. Then A is **MSP** if and only if for $x \in \mathbb{R}^n$, $Ax > 0$ implies $x > 0$.

Proof Suppose A is **MSP**, $Ax > 0$, and that $x \in \mathbb{R}$ has a nonpositive entry. If $x \geq 0$, then $x \in S_+(A)$ and by Lemma 3.5.1, A is **RSP**, a contradiction. Further, if x has a negative entry, choose $u > 0$, $u \in S_+(A)$. Then there is a convex combination $w = tu + (1-t)x, 0 < t < 1$, of u and x with $w \geq 0$ and at least one entry 0. But, as $Aw = tAu + (1-t)Ax > 0$, we again arrive at a contradiction, via Lemma 3.5.1. So, the forward implication is verified.

On the other hand, also by Lemma 3.5.1, if $Ax > 0$ implies $x > 0$, the fact that A is **SP** means that $S_+(A) \neq 0$ and no vector in $S_+(A)$ has a 0 component. So A is **MSP**, and the proof is complete. □

Note that Lemma 3.5.7 says something different from the fact claimed in Lemma 3.5.1. Now, an important characterization of square **MSP** matrices may be given.

Corollary 3.5.8 If $A \in M_n(\mathbb{R})$, then A is **MSP** if and only if A^{-1} exists and $A^{-1} \geq 0$.

Proof Suppose that A is **MSP**. It follows from Theorem 3.5.6 that A is invertible. Because A is invertible, every vector in $\mathbb{R}^n$ is in the range of A. Let $y \in \mathbb{R}^n$, $y > 0$ and let $x \in \mathbb{R}^n$ be such that $Ax = y$. By Lemma 3.5.7, $x > 0$. This means that $y > 0$ implies $A^{-1}y = x > 0$ and, because $y > 0$ is arbitrary, that $A^{-1} \geq 0$.

Conversely, suppose that $A^{-1} \geq 0$, and choose $z > 0$. Then $w = A^{-1}x > 0$, and $Aw = z > 0$, so that A is **SP**. But if $Ax = y > 0$, then $x = A^{-1}y > 0$, so that by Lemma 3.5.7, A is **MSP**, completing the proof. □

In case $m > n$, $A \in M_n(\mathbb{R})$ may still be **MSP**, in which case A will have full column rank and, thus, have *left* inverses. However, not every left inverse need to be nonnegative, unlike the square case.

Example 3.5.9 Matrix $A = \left[\begin{smallmatrix} 1 \\ 1 \end{smallmatrix}\right] \in M_{2,1}(\mathbb{R})$ is **MSP**, and $L = [2 \quad -1]$ is a left inverse that is not nonnegative. However, A does have nonnegative left inverses as well, such as $M = [1/2 \quad 1/2]$. Also, if $A = \left[\begin{smallmatrix} 3 \\ -1 \end{smallmatrix}\right]$, then A is not **SP**, but M is still a nonnegative left inverse of A. This means that non-square **MSP** matrices may not be so crisply characterized, as in Corollary 3.5.8. However, there is still a strong result.

Theorem 3.5.10 Let $A \in M_n(\mathbb{R})$ be **SP**. Then, A is **MSP** if and only if A has a nonnegative left inverse.

Proof First suppose that A is **SP** and has a nonnegative left inverse L. Suppose $x \in S_+(A)$, so that $y = Ax > 0$. Then $0 < Ly = LAx = x$, so that by Lemma 3.5.1, A is **MSP**, as no entry of a vector in $S_+(A)$ is 0.

On the other hand, suppose that A is **MSP**, so that A has left inverses by Theorem 3.5.6. Let A_j be A with the j-th column deleted, $j = 1, 2, \ldots, n$. Because A is **MSP**, none of these is **SP**. By the Theorem of the Alternative 1.4.1 and because A is **MSP**, there is a $y_j \in \mathbb{R}^m$, $y_j \geq 0, \neq 0$ with $y_j^T A_j \leq 0$. Let $L \in M_{n,m}(\mathbb{R})$ be the matrix whose j-th row is y_j^T so that $L \geq 0$. Then, the off-diagonal entries of LA are nonpositive. Let $x \in \mathbb{R}^n$, $x > 0$, be such that $Ax > 0$ (which exists as A is **MSP**), and we have $LAx > 0$. This means that $LA \in M_n(\mathbb{R})$ is a nonsingular M-matrix; see Chapter 2. Then, $B = (LA)^{-1} \geq 0$. Because $L \geq 0$, $BL \geq 0$, as well. But $(BL)A = B(LA) = (LA)^{-1}LA = I$, so that A has a nonnegative left inverse, to complete the proof. □

As was discussed in Chapter 2, there are many variations on the notion of "monotonicity" for matrices. We might call the one in Lemma 3.5.7 strong monotonicity (for not necessarily square matrices): $A \in M_{m,n}(\mathbb{R})$ is **strongly**

monotone if for $x \in \mathbb{R}^n$, $Ax > 0$ implies $x > 0$. We may summarize our understanding of **MSP** as follows.

Theorem 3.5.11 Suppose that $A \in M_{m,n}(\mathbb{R})$ is **SP**. Then, the following are equivalent:

(1) A is **MSP**.
(2) A has a nonnegative left inverse.
(3) A is strongly monotone.

In the square case, recalling that a matrix A is called monotone if $Ax \geq 0$ implies $x \geq 0$, there is the crisper statement.

Theorem 3.5.12 If $A \in M_n(\mathbb{R})$, the following are equivalent:

(1) A is **MSP**.
(2) A^{-1} exists, and $A^{-1} \geq 0$.
(3) A is monotone.

We have already discussed the sign patterns that allow or require **SP** in Section 3.3. We mention here the corresponding results for the concepts of this section: **RSP** and **MSP**. The sign patterns that allow **MSP** [JMS] are quite subtle. Some, of course, are built upon the **SP** cases.

The sign patterns that **allow RSP** are those with a $+$ in every row and for which there is a column that does not contain the only $+$ of any row.

The sign patterns that require **RSP** are those with a positive front, so that there is a column containing none of the frontal positives. To **require RSP**, a sign pattern must have a positive front, and every column must eliminate any positive front. So there must be a positive front, and each column must contain a frontal positive that is the only $+$ in its row.

The sign patterns that **allow MSP** are more complicated to describe and are not fully understood when 0s are present [JMS]. There are, as one may guess from the results of this section, closely related to sign patterns that allow a nonnegative inverse [JLR, etc.].

3.6 Linear Preservers

Suppose that $E \in M_{m,n}(\mathbb{F})$ is a class of matrices. If $L\colon M_{m,n}(\mathbb{F}) \to M_{m,n}(\mathbb{F})$ is a linear transformation such that $L(E) \subseteq E$, then L is called a **linear preserver** of E. Understanding the linear preservers of E can provide insight into class E. Often additional assumptions are made to obtain definitive results; for example, **onto** ($L(E) = E$), **invertible** (L is invertible), L of a special form (e.g., $L(x) = RXS$, $R \in M_n(\mathbb{F})$ and $S \in M_m(\mathbb{F})$ invertible).

Here, we are interested in the linear preservers of the class **SP**, and of **MSP**, in $M_{m,n}(\mathbb{F})$, in several senses. We will be able to use several of the results already developed. We first consider **SP** preservers of the form $L(X)$: $M_{m,n}(\mathbb{R}) \to M_{m,n}(\mathbb{R})$ with

$$L(X) = RXS \quad \text{for fixed} \quad R \in M_m(\mathbb{R}) \quad \text{and} \quad S \in M_n(\mathbb{R}). \tag{3.6.1}$$

We already know that if R is row positive and S is monotone, then such a transformation preserves **SP** (Theorem 3.1.5). Are there other possibilities? First is a theorem from [DGJJT16]; another proof is given in [CKS18a].

Theorem 3.6.1 Let L be a linear operator on $M_{m,n}(\mathbb{R})$ of the form (3.6.1). Then L is an into linear preserver of **SP** if and only if either R is row positive and S is monotone, or $-R$ is row positive and $-S$ is monotone.

Proof We need only verify necessity. Note that a rank one matrix xy^T is **SP** if and only if $x > 0$ and y has at least one positive entry, or $x < 0$ and y has at least one negative entry. If $X = xy^T$, with $x > 0$ and at least one entry of y is positive, then X is **SP** and $L(X) = RXS = (Rx)(S^Ty)^T$ is **SP**, per hypothesis. Because $x > 0$ is otherwise arbitrary, this means that either $Rx > 0$ or $Rx < 0$, or that either R is row positive, or $-R$ is row positive. Assume the former. Then, S^Ty has at least one positive entry for every y that has at least one positive entry. This gives that, if S^Ty has no positive entries, then y has no positive entries, i.e., S^T is monotone. But, as S is square, S must be monotone. The case in which $-R$ is row positive similarly yields that $-S$ is monotone, completing the proof. □

Recall that a matrix $B \in M_n(\mathbb{R})$ is **monomial** if $B \geq 0$ and has exactly one nonzero (i.e., positive) entry in each row and column. This is the same as saying that it is of the form $B = DP$ with $D \in D_n^+$ and $P \in P_n$ or $B = QE$ with $Q \in P_n$ and $E \in D_n^+$. The important feature of monomial matrices is that they are the intersection of the row positive and the monotone matrices in $M_n(\mathbb{R})$.

Lemma 3.6.2 A matrix $B \in M_n(\mathbb{R})$ is monomial if and only if B is both row positive and monotone.

Proof Obviously, a monomial matrix is both row positive and inverse nonnegative. Suppose now that X is row positive and monotone. Then by Theorem 3.5.12, X is also inverse nonnegative. Let e_i denote the standard basis vectors; that is, e_i has a 1 in its i-th entry and 0s everywhere else. Because X is row positive, it must take each e_i to a nonnegative sum of the e_j; that is, $Xe_i = \sum r_{ji}e_j$ for some $r_{ji} \geq 0$. Because X is inverse nonnegative, we must have that for each e_j, there exists $v_j \geq 0$ such that $Xv_j = e_j$. Note, however, that this is impossible unless for some e_i, $Xe_i = r_{ji}e_j$. If this were not true, then Xv_j would always have some positive entry that was not the j-th entry because it cannot cancel out, due to their being no negatives involved. Thus, for every e_j there exists e_i such that $Xe_i = r_{ji}e_j$. But this is exactly the same as claiming that X has exactly one non-zero entry in each row and column, and so X is monomial. □

With this, we may characterize the onto linear **SP**-preservers.

Theorem 3.6.3 A linear operator on $M_{m,n}(\mathbb{R})$ of the form (3.6.1) is an onto preserver of **SP** if and only if both R and S are monomial or $-R$ and $-S$ are monomial.

Proof Suppose that L is an onto preserver of **SP**. First, note that the set of **SP** matrices contains a basis for $M_{m,n}$. We can see this by noting that every matrix that has a 2 in one entry and a 1 in every other entry is **SP**, so each is contained in the image of L. These matrices form a basis for $M_{m,n}$. Thus, L is an invertible map, which means that R and S must be invertible, and $L^{-1}(A) = R^{-1}AS^{-1}$. Because L is an onto preserver of **SP**, both it and its inverse are into preservers of **SP**, so R and S must be both row positive and inverse nonnegative (or $-R$ and $-S$), and thus monomial. Conversely, if R and S are both monomial, then $L(A) = RAS$ is an into preserver of **SP**, as is $L^{-1}(A) = R^{-1}AS^{-1}$, so L is an onto preserver of **SP**. □

Now that we have characterized linear **SP**-preservers, both into and onto, of the form (3.6.1), we turn to linear preservers of **MSP** of this form. Notice that the onto preservers of **SP**, of this form, are also onto preservers of **MSP**. The same is not so for into preservers of **MSP**. First, we need a particular fact, not so obviously related to **SP** or **MSP**.

Lemma 3.6.4 Let $v \in \mathbb{R}^n$ and $w \in \mathbb{R}^m$ be nonzero vectors and let $n > m$. If v has both a positive entry and a negative entry, then there exists a matrix $B \in M_{m,n}(\mathbb{R})$ of full row rank such that $B \geq 0$ and $Bv = w$. The same holds if $0 \neq v \geq 0$ and $w > 0$.

Proof First, suppose that v has a negative entry. Without loss of generality, suppose that the first k entries of v are positive, and the last $n-k$ are nonnegative. Denote by $w_i \in \mathbb{R}^m$ the vector with 1 for the first $m-i+1$ entries, and 0 for every other entry. Set $Be_i = r_i w_i$ for $1 \leq i \leq m$, where $e_i \in \mathbb{R}^n$ is the i-th standard basis vector, and $r_i = 1$ for $i > 1$, and is yet to be determined for $i = 1$. Note that this forces B to have full row rank. Now, set $Be_i = 0$ for $m < i < n$. Consider the matrix B, where we complete this construction by picking $Be_n = 0$. Choose r_1 large enough that $Bv > w$, then set $Be_n = \frac{1}{v_n}(w - Bv)$, where v_n is the last entry of v. Because $v_n < 0$, this means that $Be_n \geq 0$, and we have that $Bv = w$.

Now, suppose that $v \geq 0$. Similar to the above construction, set $Be_i = r_i w_i$ for $1 < i \leq m$, and set $Be_i = 0$ for $i > m$. Let B be the end of this construction, supposing that we set $Be_1 = 0$. Take the r_i small enough that $Bv < w$, then set $Be_1 = \frac{1}{v_1}(w - Bv)$, so that $Bv = w$. □

Recall that it is necessary, for $B \in M_{m,n}(\mathbb{R})$ to be **MSP**, that $m \geq n$. It is convenient to consider linear **MSP** preservers of the form (3.6.1) in the two possible cases: $m > n$ and $m = n$.

The first case is $m > n$:

Theorem 3.6.5 Let L be a linear operator on $M_{m,n}(\mathbb{R})$, with $m > n > 1$, of form (2.6.1). Then

(i) L is an into preserver of **MSP** if and only if R is monomial and S is monotone, or $-R$ is monomial and $-S$ is monotone;

(ii) L is an onto preserver of **MSP** if and only if R and S are monomial, or $-R$ and $-S$ are monomial.

Proof

(i) First, recall that A is **MSP** if and only if A is **SP** and A has a nonnegative left inverse. Now, suppose that A is **MSP** and let B be a nonnegative left inverse for A. Then if R is monomial and S is inverse nonnegative, $S^{-1}BR^{-1}$ is nonnegative because each matrix is nonnegative. Moreover, this forms a left inverse for RAS. Because R is row positive and S is inverse nonnegative, we also know that RAS is **SP**. Thus, RAS is **MSP**. If S is also taken to be monomial, then we can obviously get every **MSP** matrix B in the image of L because we can simply set $A = R^{-1}BS^{-1}$, which is **MSP** if B is, so that $L(A) = B$. Conversely, suppose that $L(A)$ preserves **MSP** in the into sense. First, note that R and S must be invertible. If they were not, then RAS would not have full column rank, and therefore could not have a left inverse. Now, suppose that neither R nor $-R$ is inverse nonnegative. Then there exists some vector $v \not\geq 0$ and $v \not\leq 0$ such that $Rv \geq 0$. Let $w \not\geq 0$, and consider Sw. Let $B \geq 0$ be an n-by-m matrix such that $Bv = Sw$. Such a matrix must exist, by the previous lemma. Pick a basis $\{v_i\}$ for $\mathbb{R}^m$, starting with $v_1 = v$ and $v_2 > 0$, and let $z_i = Bv_i$. Note that $z_2 > 0$ because $B \geq 0$, with full row rank. Pick an m-by-n matrix A so that $Az_i = v_i$. Then A is a right inverse of B, and A is **SP** because $Az_2 = v_2$. Thus, A is **MSP**. But, $RASw = RABv = RAz_1 = Rv \geq 0$, while $w \not\geq 0$. Thus, RAS is not **MSP**, and so R or $-R$ must be inverse nonnegative.

(ii) Suppose that R is inverse nonnegative, and suppose that S was not inverse nonnegative. Then there exists some $u \geq 0$ such that $S^{-1}u = w \not\geq 0$. Set $z \geq 0$ such that $Rz \geq 0$. Such a vector exists because R is inverse nonnegative. Then choose some n-by-m matrix $B \geq 0$ with full row rank such that $Bz = u$. Note that such a matrix must exist, by the previous lemma. Now, pick a basis $\{u_i\}$ for $\mathbb{R}^n$ beginning with $u_1 = u$. Because B has full row rank, there exist linearly independent vectors z_i such that $Bz_i = u_i$, with $z_1 = z$. Set A to be an m-by-n matrix such that $Au_i = z_i$. Then A is a right inverse of B and A is **SP** because $Au = z$. Thus, A is **MSP**. But, $RASw = RAu = Rz \geq 0$, despite the fact that $w \not> 0$. Thus, RAS cannot be **MSP**, so S must be inverse nonnegative. A similar proof shows that if $-R$ is inverse nonnegative, so is $-S$. Finally, we must show that R is row positive, which comes down to just showing that $R \geq 0$ because R is invertible. Suppose that R had a negative entry. Let it be the (i,j) entry. Choose an m-by-n matrix A such that, except for the j-th row, every row has exactly one entry that is a 1 and all others 0, and every column has at least one 1 entry. This is possible, due to the fact that $m > n$. If we then choose the j-th row of A to be positive, A will be **MSP** because $A \geq 0$ with a positive entry in each row, making it **SP**, and no column-deleted submatrix could be **SP** because it would then have a zero row. Note that by making each entry of the j-th row of A large enough, we can force the i-th row of XA to be any vector $v < 0$, as long as the entries of v are large enough. Now, pick a vector $w < 0$ such that $wS < 0$. Such a vector must exist because R is inverse nonnegative, and thus so is S, which means S is **SP**. Pick A so that $v = rw$ for some large enough scalar r. Then RAS will have a negative i-th row, and thus cannot be **SP**. Therefore, R must be row positive, and so R is monomial. Now, suppose that $L(A) = RAS$ is an onto linear preserver of **MSP**. L must be invertible as shown above, so $L^{-1}(A) = R^{-1}AS^{-1}$, and $L(A) = RAS$ are both into linear preservers. That means that R is monomial and S is both inverse nonnegative and row positive, and thus also monomial. □

Corollary 3.6.6 Let $C \in M_{r,m}(\mathbb{R})$, $B \in M_{n,s}(\mathbb{R})$. Then CA (AB) is **SP**, for every $A \in M_{m,n}(\mathbb{R})$ that is **SP** if and only if C is row positive (B is monotone).

The second case is $m = n$:

The requirement that R be monomial is not necessary. In this case, R need only be monotone, because of Corollary 3.5.8 (**MSP** is equivalent to inverse nonnegativity, which is monotonicity, and monotonicity is closed under product). This should be compared with the non-square case. Preserving **SP** will occur automatically. This is revealed in the proof of Theorem 3.6.5 when A, with one row deleted is asked to be **MSP**. This could not happen when $m = n$. Thus, we have

Theorem 3.6.7 Let L be a linear operator on $M_n(\mathbb{R})$ of the form (3.6.1). Then

(i) L is an into preserver of **MSP** if and only if R and S are monotone, or $-R$ and $-S$ are monotone;

(ii) L is an onto preserver of **MSP** if and only if R and S are monomial, or $-R$ and $-S$ are monomial.

If we only ask that $L\colon M_{m,n}(\mathbb{R}) \to M_{m,n}(\mathbb{R})$ be an onto linear preserver of **SP**, with no requirement of special form, it turns out that the special form (3.6.1) is necessary for linear, *onto*, preservation of **SP**. But, this requires substantial proof. We begin with some needed lemmas.

Lemma 3.6.8 If $L\colon M_{m,n}(\mathbb{R}) \to M_{m,n}(\mathbb{R})$ is an onto linear preserver of **SP**, then $L(X)$ is **SP** if and only if X is **SP**.

Proof The result follows from the definition of an onto preserver, and the fact that the **SP** class of matrices forms a basis. □

Lemma 3.6.9 If $L\colon M_{m,n}(\mathbb{R}) \to M_{m,n}(\mathbb{R})$ is an onto linear preserver of **SP**, then L is also an onto linear preserver of **LSP**.

Proof First, note that L is an onto preserver of the set of matrices that are not left **SNP** because this is the same as the set of matrices that are **SP**, by the Theorem of the Alternative 1.4.1. But, L must be invertible because the set of **SP** matrices contains a basis for $M_{m,n}$, so L in fact must be an onto preserver of left **SNP**. L is a homeomorphism, as it and its inverse are continuous, and the set of left **SNP** matrices is the closure of the set of left **SN** matrices, so L must also be an onto preserver of left **SN**. But the set of left **SN** matrices is exactly the negative of the set of left **SP** matrices, so if one is preserved in the onto sense by a linear map, so must the other. Thus, L preserves left **SP**. □

The above reasoning can be used to show that a linear map preserving one class of matrices must also preserve others. Now that we realize that onto linear preservers of **SP** are also **LSP** preservers, we can transfer any fact about the columns of matrices after L is applied to them, to the rows, as well. To show that an onto linear preserver of **SP** preserves nonnegative matrices, as well, we need an observation about the relationship between nonnegative matrices and **SP** matrices that we could have made earlier.

Lemma 3.6.10 A matrix $A \in M_{m,n}(\mathbb{R})$ satisfies $A \geq 0$ if and only if, for every **SP** matrix $B \in M_{m,n}(\mathbb{R})$, $B + A$ is also **SP**.

Proof First, suppose A is nonnegative. Then for any $v \geq 0$, $Av \geq 0$. Thus, if $Bv > 0$, then $(A + B)v = Av + Bv > 0$, so $A + B$ is also **SP**. Conversely, suppose that A is not nonnegative. Then A has a negative entry. Suppose it to be the (i,j) entry. Pick B so that the j-th column of B is positive and all other entries are negative and larger in absolute value than any entry of A. Further, have the (i,j) entry of B smaller in absolute value than the (i,j) entry of A. Then B is **SP** because it has a positive column, but $A + B$ is not **SP** because every entry in the i-th row must be negative. □

We may now see that onto linear **SP**-preservers are onto linear preservers of the nonnegative matrices.

Lemma 3.6.11 If $L\colon M_{m,n}(\mathbb{R}) \to M_{m,n}(\mathbb{R})$ is an onto linear preserver of **SP**, then it is also an into linear preserver of nonnegativity, that is, if $A \in M_{m,n}(\mathbb{R})$ and $A \geq 0$, then $L(A) \geq 0$.

Proof Let $A \geq 0$. Then for any **SP** matrix B, $A + B$ is **SP**, so $L(A + B) = L(A) + L(B)$ is **SP**. Note also that $L(B)$ is **SP**. Further, for any **SP** matrix C there exists an **SP** matrix B such that $L(B) = C$. Thus, for any **SP** matrix C, $L(A) + C$ is **SP**. By the previous lemma, then, $L(A) \geq 0$. □

We now know that if we take the matrix E_{ij}, an element of a basis of $M_{m,n}(\mathbb{R})$, and apply a onto linear preserver L of **SP** to it, $L(E) \geq 0$. Further, any fact about the columns of $L(E_{ij})$, resulting from this, applies to its rows, as well. Then, we get a general **SP**-preserver result.

Theorem 3.6.12 Suppose $L\colon M_{m,n}(\mathbb{R}) \to M_{m,n}(\mathbb{R})$. Then, L is an onto linear preserver of **SP** if and only if L is of the form (3.6.1) for some monomial matrices R and S.

Proof We already know the reverse implication. Now suppose L is an onto linear preserver of **SP**. First, we know that $L(E_{ij}) \geq 0$, by the previous lemma and the invertibility of L. Suppose that $L(E_{ij})$ has positive entries in more than one row. Note that $L\left(\sum_i E_{ij}\right)$ must be **SP** because $\sum_i E_{ij}$ is **SP** (this is the matrix with 1s in the j-th row and 0s elsewhere). However, $L\left(\sum_{i\neq k} E_{ij}\right)$ cannot be **SP** for any k because $\sum_{i\neq k} E_{ij}$ is not **SP**, and L is an onto preserver of **SP**. Now, each $L(E_{ij})$ must have a positive entry in at least one row. If any had positive entries in two rows, there would be one E_{kj} that only had positive entries in rows that another E_{ij} already has a positive entry. But then, $L\left(\sum_{i\neq k} E_{ij}\right)$ would be **SP** because every row would have a positive entry, and no entry would be negative, so the sum of each column would be positive. Thus, each $L(E_{ij})$ can only have positive entries in one row. By the fact that L is also a preserver of left **SP**, each $L(E_{ij})$ can only have positive entries in one column as well. Thus, each $L(E_{ij})$ is a matrix with exactly one positive entry, which cannot be shared by any other $L(E_{kl})$, due to the invertibility of L.

Next, suppose that the $L(E_{ij})$ for fixed j are sent to matrices with the positive entry in different columns. Then consider a matrix A with a small positive

j-th column, and large negative entries everywhere else. If the entries in the j-th column of A are not sent to the same column of $L(A)$, then we will end up with a matrix whose positive entries are not all in the same column. Because we can make the negative entries as large as we wish, we can force $L(A)$ to not be **SP**, even though A was. Thus, all entries in the same column of A must be sent to the same column of $L(A)$. Again, this must also be true for rows.

The above now forces L to be a composition of Hadamard multiplication by some positive matrix, and permutation of rows and columns. Because permutation matrices are all monomial, we can assume that L is just Hadamard multiplication by some positive matrix, so that $L(E_{ij}) = r_{ij}E_{ij}$.

Consider the matrix

$$A = \begin{bmatrix} 0 & \cdots & 0 & 1 & 1 & 0 & \cdots & 0 \\ \vdots & & \vdots & \vdots & \vdots & \vdots & & \vdots \\ 0 & \cdots & 0 & 1 & 1 & 0 & \cdots & 0 \\ 0 & \cdots & 0 & 1 & -x & 0 & \cdots & 0 \\ 0 & \cdots & 0 & -x & 1 & 0 & \cdots & 0 \\ 0 & \cdots & 0 & 1 & 1 & 0 & \cdots & 0 \\ \vdots & & \vdots & \vdots & \vdots & \vdots & & \vdots \\ 0 & \cdots & 0 & 1 & 1 & 0 & \cdots & 0 \end{bmatrix}.$$

Note particularly that A is **SP** if and only if $x < 1$. Then

$$L(A) = \begin{bmatrix} \vdots & \vdots & \vdots & \vdots & \vdots & \vdots \\ 0 & \cdots & r_{i,j} & -xr_{i,j+1} & \cdots & 0 \\ 0 & \cdots & -xr_{i+1,j} & r_{i+1,j+1} & \cdots & 0 \\ \vdots & \vdots & \vdots & \vdots & \vdots & \vdots \end{bmatrix}.$$

$L(A)$ is **SP** if and only if A is **SP**, so $L(A)$ is **SP** if and only if $x < 1$. Because all the other columns are zero besides j and $j+1$, and all other rows of j and $j+1$ are positive except i and $i+1$, whether $L(A)$ is **SP** depends entirely on those four entries. Setting $x = 1$, we see that we must have $r_{i,j} = ar_{i,j+1}$ and $r_{i+1,j} = ar_{i+1,j+1}$. But we can repeat this for every pair of columns and rows to find the number we multiply an entry of some column by is in a fixed ratio with the number we multiply the entry of a different column in the same row by. This is just multiplication on the right by a positive diagonal matrix. Again, we can apply the fact that L must be left-**SP** preserving to show that the ratio of rows must also be in a fixed ratio, and so L is just multiplication on the left and right by positive diagonal matrices. But these are also monomial matrices, so any onto linear preserver of **SP** can be written as $L(A) = RAS$ for monomial R and S. $\square$

For into linear preservers of **SP**, the special form (3.6.1) need not follow.

Example 3.6.13 Consider the linear map $L\colon M_2(\mathbb{R}) \to M_2(\mathbb{R})$ given by

$$L\left(\begin{bmatrix} a & b \\ c & d \end{bmatrix}\right) = \begin{bmatrix} a & b \\ a & b-a \end{bmatrix}.$$

This map preserves **SP** and is neither of the form (3.6.1) nor invertible. For the argument A to be **SP**, either a or b must be positive. If $a > 0$, the first column of $L(A)$ is positive, so that $L(A)$ is **SP**. If $b > 0$ and $a \leq 0$, then the second column of $L(A)$ is positive, and $L(A)$ is **SP**. Thus, L preserves **SP**. Because

$$L\left(\begin{bmatrix} 1 & 0 \\ 0 & 0 \end{bmatrix}\right) = \begin{bmatrix} 1 & 0 \\ 1 & -1 \end{bmatrix},$$

L can increase rank and so cannot be of form (3.6.1).

More generally, consider the map on $M_{m,n}(\mathbb{R})$ given by

$$L([a_{ij}]) = \begin{bmatrix} a_{11} & a_{12} & \cdots & a_{1n} \\ a_{11} & a_{12} - a_{11} & \cdots & a_{1n} - \sum\limits_{k=1}^{n-1} a_{1k} \\ \vdots & \vdots & & \vdots \\ a_{11} & a_{12} - a_{11} & \cdots & a_{1n} - \sum\limits_{k=1}^{n-1} a_{1k} \end{bmatrix}.$$

Because any **SP** matrix has a positive entry in the first row, $L[a - ij]$ has a positive column, so that **SP** is preserved. But, as L can increase rank, L is not of form (3.6.1).

Remark 3.6.14 The maps in the example above are not invertible. It is an open question what is the least strong regularity condition on an into **SP** preserver that it be of form (3.6.1), or what all into linear **SP** preservers are.

3.7 Geometric Mapping Properties of SP Matrices

One would not expect that many strong properties of entry-wise nonnegative matrices generalize to **SP** matrices. There are, however, some cone-theoretic and Perron–Frobenius-type implications of semipositivity. This section contains such connections of **SP** matrices to cones of nonnegative vectors and cone invariance. More results of this type, including conditions under which an **SP** matrix has a positive eigenvalue, or leaves a proper cone invariant can be found in [Tsa16] and [ST18].

3.7.1 Preliminaries

We will consider the following set, which is the closure of $S_+(A)$.

Definition 3.7.1 For $A \in M_{m,n}(\mathbb{R})$, let

$$K_+(A) = \{x \in \mathbb{R}^n : x \geq 0 \text{ and } Ax \geq 0\}.$$

Note that A is **SP** if and only if $AK_+(A)$ contains a positive vector, in which case we refer to $K_+(A)$ as the **semipositive cone** of A.

Many of the results on $K_+(A)$ herein are stated and indeed hold for arbitrary $A \in M_{m,n}(\mathbb{R})$. Our interest, however, is in semipositive matrices. The goal is to compute and study $K_+(A)$ as a convex cone in $\mathbb{R}^n$. For that purpose, we review below some basic material on generalized inverses and cones.

The **Moore–Penrose** inverse of $A \in M_{m,n}(\mathbb{R})$ is denoted by $A^\dagger$, and the **group inverse** of A is denoted by $A^\#$. Their defining properties are, respectively,

$$AA^\dagger A = A,\ A^\dagger AA^\dagger = A^\dagger,\ (AA^\dagger)^T = AA^\dagger, (A^\dagger A)^T = A^\dagger A,$$

and

$$AA^\# A = A,\ A^\# AA^\# = A^\#, AA^\# = A^\# A.$$

We let the **range** of A be denoted by $R(A)$ and its **null space** by $\text{Null}(A)$. While the Moore–Penrose inverse exists for all matrices A, the group inverse of a square matrix $A \in M_n(\mathbb{R})$ exists if and only if $\text{rank}(A) = \text{rank}(A^2)$ (equivalently, $\text{Null}(A) = \text{Null}(A^2)$). It is known that $A^\#$ exists if and only if the $R(A)$ and $\text{Null}(A)$ are complementary subspaces of $\mathbb{R}^n$.

We call $A \in M_n(\mathbb{R})$ **range symmetric** if $R(A) = R(A^T)$ or, equivalently, if $A^\dagger = A^\#$. The following will be used in some of the proofs below: $R(A^\dagger) = R(A^T)$, $\text{Null}(A^\dagger) = \text{Null}(A^T)$, $R(A^\#) = R(A)$, and $\text{Null}(A^\#) = \text{Null}(A)$.

Recall that the topological interior of $\mathbb{R}^n_+$ comprises all positive vectors in $\mathbb{R}^n$ and is denoted by $\text{int}\,\mathbb{R}^n_+$. We use $x > 0$ and $x \in \text{int}\,\mathbb{R}^n_+$ interchangeably. The following geometric concepts will also be used in the sequel.

The **dual** of a set $S \subseteq \mathbb{R}^n$ is $S^* = \{z \in \mathbb{R}^n : z^T y \geq 0 \text{ for all } y \in S\}$.

A nonempty convex set $K \subseteq \mathbb{R}^n$ is said to be a **cone** if $\alpha K \subseteq K$ for all $\alpha \geq 0$. A cone K is called a **proper cone** if it is (i) *closed* (in the Euclidean space $\mathbb{R}^n$),

(ii) *pointed* (i.e., $K \cap (-K) = \{0\}$), and (iii) *solid* (i.e., the topological interior of K, $\operatorname{int} K$, is nonempty).

A **polyhedral cone** $K \subseteq \mathbb{R}^m$ is a cone consisting of all nonnegative linear combinations of a finite set of vectors in $\mathbb{R}^m$, which are called the **generators** of K. Thus, K is polyhedral if and only if $K = X\mathbb{R}^n_+$ for some $X \in M_{m,n}(\mathbb{R})$; when $m = n$ and X is invertible, $K = X\mathbb{R}^n_+$ is called a **simplicial cone** in $\mathbb{R}^n_+$. Note that simplicial cones in $\mathbb{R}^n$ are proper cones.

A cone K is called an **acute cone** if $p^T q \geq 0$ for all $p, q \in K$. In terms of its dual, acuteness is equivalent to the inclusion $K \subseteq K^*$. A dual notion is that of obtuseness; K is called an **obtuse cone** if $K^* \subseteq K$. K is called a **self-dual cone** if it is both acute and obtuse. $\mathbb{R}^n_+$ is an example of a self-dual cone.

We will have occasion to use the following results. The first result on the consistency of linear equations is quite well known; see, e.g., [B-IG03].

Lemma 3.7.2 Let $A \in M_{m,n}(\mathbb{R})$ with $b \in \mathbb{R}^m$. Then the system of linear equations $Ax = b$ has a solution if and only if $AA^\dagger b = b$. In such a case, the general solution is given by $x = A^\dagger b + z$, where $z \in \operatorname{Null}(A)$.

A version of the separating hyperplane theorem, which will be used below, is recalled next. For its proof, see [Man69].

Theorem 3.7.3 Let $K \subseteq \mathbb{R}^n$ be a closed convex set and $b \notin K$. Then there exists $c \in \mathbb{R}^n$ and a real number α such that

$$c^T b < \alpha \leq c^T x \text{ for all } x \in K.$$

3.7.2 Cones Associated with SP Matrices

First is a fundamental factorization of **SP** matrices into the product of a positive and an inverse positive matrix.

Theorem 3.7.4 $A \in M_{m,n}(\mathbb{R})$ is **SP** if and only if there exist positive matrices $X \in M_n(\mathbb{R})$ and $Y \in M_{m,n}(\mathbb{R})$ such that X is invertible and $A = YX^{-1}$.

Proof Let A be semipositive, i.e., there exist positive vectors $x \in \operatorname{int}\mathbb{R}^n_+$ and $y \in \operatorname{int}\mathbb{R}^m_+$ such that $Ax = y$. Define the matrices

$$X = xe^T + \epsilon I \in M_n(\mathbb{R}), \quad Y = ye^T + \epsilon A \in M_{m,n}(\mathbb{R}),$$

where e generically denotes the all-ones vector of appropriate size, and $\epsilon > 0$ is chosen sufficiently small to have $Y > 0$. Then the result follows from the facts that $AX = Y$ and that $X > 0$ is invertible because its eigenvalues are ϵ and $e^T x + \epsilon$. For the converse, assume there are positive matrices $X \in M_n(\mathbb{R})$ and

$Y \in M_{m,n}(\mathbb{R})$ such that $A = YX^{-1}$. Let $u \in \operatorname{int}\mathbb{R}^n_+$ and set $v = Xu \in \operatorname{int}\mathbb{R}^n_+$. Then $Av = YX^{-1}v = Yu \in \operatorname{int}\mathbb{R}^m_+$, showing that A is **SP**. □

The following theorem shows that **SP** matrices act like nonnegative matrices on polyhedral subcones of $\mathbb{R}^n_+$.

Theorem 3.7.5 $A \in M_{m,n}(\mathbb{R})$ is **SP** if and only if there exist proper polyhedral cone $K_1 \subseteq \mathbb{R}^n_+$ and polyhedral cone $K_2 \subseteq \operatorname{int}\mathbb{R}^m_+ \cup \{0\}$ such that $AK_1 = K_2$.

Proof Let A be **SP**. Consider the matrices X, Y in the proof of Theorem 3.7.4 such that $A = YX^{-1}$ and let

$$K_1 = X\,\mathbb{R}^n_+ \quad \text{and} \quad K_2 = Y\,\mathbb{R}^n_+.$$

Because X is positive and invertible, K_1 is simplicial and thus a proper cone in $\mathbb{R}^n_+$. Because $Y > 0$, K_2 is a polyhedral cone in $\operatorname{int}\mathbb{R}^m_+ \cup \{0\}$. We also have that $AK_1 = YX^{-1}X\mathbb{R}^+ = K_2$. For the converse, suppose there exists a proper cone $K_1 \subseteq \mathbb{R}^n_+$ and a polyhedral cone $K_2 \subseteq \operatorname{int}\mathbb{R}^m_+ \cup \{0\}$ such that $AK_1 = K_2$. As $K_1 \subseteq \mathbb{R}^n_+$ is proper, it is solid and so there is $x \in \operatorname{int}K_1 \subseteq \operatorname{int}\mathbb{R}^n_+$. It follows that $Ax \in K_2 \setminus \{0\} \subseteq \operatorname{int}\mathbb{R}^m_+$. Thus A is **SP**. □

Next, we turn our attention to $K_+(A)$ and its dual, $K_+(A)^*$.

Theorem 3.7.6 Let $A \in M_{m,n}(\mathbb{R})$. Then $K_+(A)^* = A^T(\mathbb{R}^m_+) + \mathbb{R}^n_+$.

Proof Let $y = A^T u + v$, where $u \in \mathbb{R}^m_+$, $v \in \mathbb{R}^n_+$, and let $x \in K_+(A)$. Then $x \geq 0$ and $Ax \geq 0$. We have

$$x^T y = x^T \left(A^T u + v\right) = (Ax)^T u + x^T v \geq 0,$$

so that $A^T(\mathbb{R}^m_+) + \mathbb{R}^n_+ \subseteq K_+(A)^*$. Conversely, let $y \in K_+(A)^*$. Suppose that $y \notin A^T(\mathbb{R}^m_+) + \mathbb{R}^n_+$. Then by Theorem 3.7.3, there exists a vector p and a number α such that

$$p^T y < \alpha \leq p^T \left(A^T u + v\right) \text{ for all } u \in \mathbb{R}^m_+ \text{ and } v \in \mathbb{R}^n_+.$$

By setting $u = 0$ and $v = 0$, we then have $\alpha \leq 0$. Replacing u by tu and v by tv, for $t > 0$, one has

$$\alpha \leq tp^T \left(A^T u + v\right) \text{ for all } u \in \mathbb{R}^m_+ \text{ and } v \in \mathbb{R}^n_+.$$

Then

$$\frac{\alpha}{t} \leq p^T \left(A^T u + v\right) \text{ for all } u \in \mathbb{R}^m_+ \textit{ and } v \in \mathbb{R}^n_+.$$

Letting $t \to \infty$, we have

$$p^T y < \alpha \leq 0 \leq p^T \left(A^T u + v\right) \text{ for all } u \in \mathbb{R}^m_+ \text{ and } v \in \mathbb{R}^n_+.$$

By setting $u = 0$ and $v = 0$ separately, we obtain

$$p^T v \geq \text{ for all } v \geq 0 \text{ and } (Ap)^T u \geq 0 \text{ for all } u \geq 0,$$

showing that $p \geq 0$ and $Ap \geq 0$, i.e., $p \in K_+(A)$. Because $p^T y < 0$, we arrive at a contradiction to $y \in K_+(A)^*$, completing the proof of the reverse inclusion. □

Corollary 3.7.7 Let $A \in M_{m,n}(\mathbb{R})$ be **SP**. Then $K_+(A)$ is a proper polyhedral cone in $\mathbb{R}^n$.

Proof First, $K_+(A)$ is clearly a convex set that is closed under nonnegative scaling. Because A is semipositive, $K_+(A)$ contains a positive vector. That is, $K_+(A)$ is a nontrivial cone in $\mathbb{R}^n$. Let now $x \in \operatorname{int} \mathbb{R}^n_+$ such that $Ax \in \operatorname{int} \mathbb{R}^m_+$ and let X, Y, K_1, K_2 as in the proof of Theorem 3.7.5, so that $AK_1 = K_2$. Hence, K_1 is a simplicial and consequently a proper polyhedral cone that is contained in $K_+(A)$. It follows that $K_+(A)$ contains a solid cone and therefore $K_+(A)$ is solid. Also, $K_+(A) \subseteq \mathbb{R}^n_+$ and consequently K is pointed. Last, $K_+(A)$ is clearly a closed set. Thus, $K_+(A)$ is proper cone in $\mathbb{R}^n$. Next, recall that any nonempty subset of $\mathbb{R}^n$ is a polyhedral cone if and only if its dual set is a polyhedral cone; see e.g., [BP94, chapter 1, theorem (2.5)(c)]. By Theorem 3.7.6, $K_+(A)^* = [A^T \; I_n] \mathbb{R}^n_+$, that is, $K_+(A)^*$ is the cone generated by the n columns of A^T and the columns of the n-by-n identity matrix. It follows that K_A^*, and thus $K_+(A)$, are polyhedral cones. □

In what follows, we determine a necessary and sufficient condition for $K_+(A)$ to be self-dual.

Corollary 3.7.8 Let $A \in M_{m,n}(\mathbb{R})$ be **SP**. Then the cone $K_+(A)$ is acute. $K_+(A)$ is obtuse if and only if $A \geq 0$. $K_+(A)$ is a self-dual cone if and only if $A \geq 0$.

Proof Let $x \in K_+(A)$ be fixed and $y \in K_+(A)$ be arbitrary. Then $x \geq 0$ and $y \geq 0$ so that $x^T y \geq 0$, showing that $x \in K_+(A)^*$. Thus $K_+(A)$ is an acute cone. Next, let $A \geq 0$. Then $K_+(A) = \mathbb{R}^n_+$ and so $K_+(A)^* = \mathbb{R}^n_+$. Thus $K_+(A)^* \subseteq K_+(A)$, i.e., $K_+(A)$ is obtuse. Conversely, let $K_+(A)$ be obtuse, i.e., $K_+(A)^* \subseteq K_+(A)$. Then

$$A^T(\mathbb{R}^m_+) + \mathbb{R}^n_+ \subseteq K_+(A) \subseteq \mathbb{R}^n_+,$$

so that for every $u \in \mathbb{R}^m_+$ and for every $v \in \mathbb{R}^n_+$, one has $A^T u + v \geq 0$. By setting $v = 0$, we then have $A^T u \geq 0$ for every $u \in \mathbb{R}^m_+$. This means that $A^T \geq 0$, i.e., $A \geq 0$. This proves the second statement. □

In [Tsa16, remark 4.6] it is remarked that if a matrix B maps $K_+(A)$ into the nonnegative orthant, then the inclusion $K_+(A) \subseteq K_+(B)$ holds. In the next result, among other sufficient conditions that guarantee such an inclusion, we consider the converse question.

Theorem 3.7.9 Let $A, B \in M_{m,n}(\mathbb{R})$ with $\text{Null}(A) \subseteq \text{Null}(B)$. Consider the following statements:

(a) $BA^\dagger \geq 0$.
(b) $Ax \geq 0 \Rightarrow Bx \geq 0$.
(c) $B(K_+(A)) \subseteq \mathbb{R}^n_+$.
(d) $K_+(A) \subseteq K_+(B)$.
(e) There exists $W \geq 0$ such that $B = WA$.

Then (a) $\Rightarrow$ (b) $\Rightarrow$ (c) $\Rightarrow$ (d) $\Rightarrow$ (e). Suppose further, that $AA^\dagger \geq 0$. Then all the statements are equivalent.

Proof
(a) $\Rightarrow$ (b) Let $y = Ax \geq 0$. Then $x = A^\dagger y + z$, for some $z \in \text{Null}(A) \subseteq \text{Null}(B)$. Thus, $Bx = BA^\dagger y \geq 0$ because $BA^\dagger \geq 0$.

(b) $\Rightarrow$ (c) Let $x \in K_+(A)$ so that $x \geq 0$ and $Ax \geq 0$. Then $Bx \geq 0$ and so (c) holds.

(c) $\Rightarrow$ (d) Let $x \in K_+(A)$ so that $x \geq 0$. Then $Bx \geq 0$ and so $x \in K_+(B)$.

(d) $\Rightarrow$ (e) Let $K_+(A) \subseteq K_+(B)$. Then $(K_+(B))^* \subseteq (K_+(A))^*$ so that by Theorem 3.7.6, we have the following inclusion:

$$B^T(\mathbb{R}^m_+) + \mathbb{R}^n_+ \subseteq A^T(\mathbb{R}^m_+) + \mathbb{R}^n_+.$$

In particular, for every $u \in \mathbb{R}^m_+$, there exists $v \in \mathbb{R}^m_+$ such that $B^T u = A^T v$. By substituting the standard basis elements of $\mathbb{R}^m_+$ for the vector u, it then follows that there exists a matrix V such that $V \geq 0$ and $B^T = A^T V$. By setting $V = W^T$, we obtain (e). Let us now assume that $AA^\dagger \geq 0$. Then $BA^\dagger = WAA^\dagger \geq 0$, showing that (e) $\Rightarrow$ (a) and thus all the statements are equivalent. □

There is a nice relationship between $K_+(A)$ and $K_+(A^{-1})$ when A is invertible. We obtain this as a consequence of the next general result involving the Moore–Penrose inverse.

Theorem 3.7.10 Let $A \in M_n(\mathbb{R})$ be range symmetric and $AA^\dagger \geq 0$. Then

$$A(K_+(A)) + \text{Null}(A^T) = \{y \colon y \in \mathbb{R}^n_+ + \text{Null}(A^T) \text{ and } A^\dagger y \geq 0\}.$$

Proof Because A is range symmetric, $A^\dagger$ commutes with A. Also, $\text{Null}\left(A^T\right) = \text{Null}(A^\dagger)$. Let $y = Ax + z$, where $x \in K_+(A)$ and $z \in \text{Null}\left(A^T\right)$. Then $x \geq 0$ and

$Ax \geq 0$ so that one has $y \in \mathbb{R}^n_+ + \text{Null}\left(A^T\right)$. Also, $A^\dagger y = A^\dagger Ax = AA^\dagger x \geq 0$, proving the inclusion

$$A(K_+(A)) + \text{Null}\left(A^T\right) \subseteq \{y\colon y \in \mathbb{R}^n_+ + \text{Null}\left(A^T\right) \text{ and } A^\dagger y \geq 0\}.$$

On the other hand, suppose that $y = u + v$, where $u \geq 0$ and $v \in \text{Null}\left(A^T\right)$. Also, let $A^\dagger y \geq 0$. On letting

$$w = A^\dagger u = A^\dagger y \geq 0,$$

we have that $u = Aw + z$ for some $z \in \text{Null}\left(A^T\right)$. Also,

$$Aw = AA^\dagger u = A^\dagger Au \geq 0,$$

so that $w \in K_+(A)$. We then have $y = u + v = Aw + (z + v) \in A(K_+(A)) + \text{Null}\left(A^T\right)$, proving the inclusion in the reverse direction. □

The following is an easy consequence of the result above, applied to an invertible matrix.

Corollary 3.7.11 Let $A \in M_n(\mathbb{R})$ be invertible. Then $A(K_+(A)) = K_+(A^{-1})$.

In the next result, we present a class of matrices that map the cone $K_+(A)$ into the cone $K_+(B)$.

Lemma 3.7.12 Let $A, B \in M_{m,n}(\mathbb{R})$ such that $R(A) \subseteq R(B)$ and $B^\dagger \geq 0$. Then $B^\dagger A(K_+(A)) \subseteq K_+(B)$. In particular, if $A^\dagger \geq 0$, then the cone $K_+(A)$ is invariant under $A^\dagger A$.

Proof Let $x \in K_+(A)$ so that $x \geq 0$ and $Ax \geq 0$. Set $y = B^\dagger Ax$. Then $y \geq 0$ and $By = BB^\dagger Ax = Ax \geq 0$, where we have made use of the fact that $BB^\dagger A = A$ because $R(A) \subseteq R(B)$. The second part is a simple consequence of the first part. □

Remark 3.7.13 Let $A = \begin{bmatrix} 1 & 0 \\ 0 & 0 \end{bmatrix}$ so that $K_+(A) = \mathbb{R}^n_+$. Let $B = \begin{bmatrix} -1 & -1 \\ 0 & 0 \end{bmatrix}$. Then $R(A) \subseteq R(B)$. Note that $B^\dagger = \frac{1}{2}\begin{bmatrix} -1 & 0 \\ -1 & 0 \end{bmatrix} \ngeq 0$. Let $x^0 = [2 \quad 0]^T$. Then $x^0 \in K_+(A)$. However,

$$B^\dagger Ax^0 = [-1 \quad -1]^T \ngeq 0.$$

Hence $B^\dagger A(K_+(A)) \nsubseteq K_+(B)$, showing that the condition $B^\dagger \geq 0$ is indispensable in the result above.

Let $A = \begin{bmatrix} 0 & 0 & 0 \\ 1 & 0 & 0 \\ 0 & 1 & 0 \end{bmatrix}$ so that $K_+(A) = \mathbb{R}^n_+$. Let $B = \frac{1}{3}\begin{bmatrix} 1 & 0 & 0 \\ 0 & 1 & 0 \\ 1 & 1 & 0 \end{bmatrix}$. Then $[0 \;\; 1 \;\; 0]^T \in R(A)$ and does not belong to $R(B)$. It may be verified that $B^\dagger = \begin{bmatrix} 2 & -1 & 1 \\ -1 & 2 & 1 \\ 0 & 0 & 0 \end{bmatrix}$. If $x^0 = [1 \;\; 0 \;\; 0]^T$, then $x^0 \in K_+(A)$ but $B^\dagger A x^0 \not\geq 0$, so that $B^\dagger A(K_+(A)) \not\subseteq K_+(B)$. This shows that the range inclusion condition $R(A) \subseteq R(B)$ cannot be removed.

We continue with a representation of the semipositive cone $K_+(A)$, when $AA^\dagger \geq 0$.

Theorem 3.7.14 Let $A \in M_{m,n}(\mathbb{R})$ with $AA^\dagger \geq 0$. Then

$$K_+(A) = \mathbb{R}^n_+ \cap (A^\dagger \mathbb{R}^m_+ + \mathrm{Null}(A)).$$

Proof Let $x \in K_+(A)$ and $y = Ax \in \mathbb{R}^m_+$. Then $x \geq 0$ and, by Lemma 3.7.2, one has $x = A^\dagger y + z$ for some $z \in \mathrm{Null}(A)$, proving one-way inclusion. If $x \in \mathbb{R}^n_+ \cap (A^\dagger \mathbb{R}^m_+ + \mathrm{Null}(A))$, then $x \geq 0$ and $x = A^\dagger u + v$ for some $u \in \mathbb{R}^m_+$ and $v \in \mathrm{Null}(A)$. Then, because $AA^\dagger \geq 0$, one has $Ax = AA^\dagger u \in AA^\dagger \mathbb{R}^m_+ \subseteq \mathbb{R}^m_+$. Thus $x \in K_+(A)$, proving the reverse inclusion. □

Corollary 3.7.15 Let $A \in \mathbb{R}^n$ be invertible. Then $K_+(A) = \mathbb{R}^n_+ \cap A^{-1}(\mathbb{R}^n_+)$.

Remark 3.7.16 Let $A = \begin{bmatrix} 2 & 1 & -1 \\ -2 & -1 & 1 \end{bmatrix}$. Then $A^\dagger = \frac{1}{12}A^T$ so that $AA^\dagger = 6\begin{bmatrix} 1 & -1 \\ -1 & 1 \end{bmatrix} \not\geq 0$. Set $x^0 = A^\dagger e^1 + v$, where $e^1 = [1 \;\; 0]^T$ and $v = \frac{1}{12}[0 \;\; 1 \;\; 1]^T \in \mathrm{Null}(A)$. Then $x^0 \in \mathbb{R}^3_+ \cap (A^\dagger \mathbb{R}^2_+ + \mathrm{Null}(A))$. However, $Ax^0 = \frac{1}{2}[1 \;\; -1]^T \not\geq 0$. This shows that the representation in Theorem 3.7.14 is not valid without the assumption that $AA^\dagger \geq 0$.

3.7.3 Intervals of Semipositive Matrices

Entry-wise nonnegativity induces a natural partial order among matrices of the same size, namely, $Y \geq X$ if and only if $Y - X \geq 0$. When $Y \geq X$, one can consider the **matrix interval** $[X, Y]$ that comprises all matrices Z such that $X \leq Z \leq Y$.

We recall a result on intervals of inverse positive matrices; see Rohn [Roh87, theorem 1].

Theorem 3.7.17 $A \in M_n(\mathbb{R})$ and $B \in M_n(\mathbb{R})$ are inverse positive if and only if each $X \in [A, B]$ is inverse positive.

It is easy to observe that if $A \leq C$ and if A is semipositive, then C is semipositive. This may be paraphrased by the statement that if $A \leq C$, then $K_+(A) \subseteq K_+(C)$. It prompts us to ask the following question: Let A be semipositive. What are the possible choices for *semipositive* matrices $C \leq A$? In this regard, we will consider diagonal and rank-one perturbations of a semipositive matrix A of the form $A - D$ or $A - uv^T$, where D is a nonnegative diagonal matrix and u, v are nonnegative vectors.

A matrix interval $[C, E]$ will be referred to as a **semipositive interval** if each matrix X such that $C \leq X \leq E$ is semipositive. A similar nomenclature is adopted for minimal semipositivity.

Theorem 3.7.18 Let $A \in M_n(\mathbb{R})$ be semipositive and let $D \in M_n(\mathbb{R})$ be a nonnegative diagonal matrix with diagonal entries d_j, $j = 1, 2, \ldots n$. Then $[A - D, A]$ is a semipositive interval if and only if there exists positive vector $x \in K_+(A)$ such that for each $j = 1, 2, \ldots, n$,

$$0 \leq d_j < \frac{(Ax)_j}{x_j}.$$

Proof Let $x \in K_+(A)$ be positive and suppose that for each $j = 1, 2, \ldots, n$,

$$0 \leq d_j < \frac{(Ax)_j}{x_j}.$$

Then the j-th entry of $(A - D)x$ is

$$((A - D)x)_j = (Ax)_j - d_j x_j > (Ax)_j - \frac{(Ax)_j}{x_j} x_j = 0;$$

that is, $A - D$ is semipositive. If $C \in [A - D, A]$, then $Cx \geq (A - D)x > 0$ and so $[A - D, A]$ is a semipositive interval. Conversely, if D is a nonnegative diagonal matrix with diagonal entries d_j and $[A - D, A]$ is a semipositive interval, then there exists positive x such that $Ax > Dx \geq 0$. It follows that $x \in K_+(A)$ and $0 \leq d_j < (Ax)_j / x_j$. □

Theorem 3.7.19 Let $A \in M_n(\mathbb{R})$ be semipositive and let $D \in M_n(\mathbb{R})$ be a nonnegative diagonal matrix with diagonal entries $d_j, j = 1, 2, \ldots n$. Let

$$\alpha = \max\{x^T A x \colon \|x\|_2 = 1, x \in K_+(A)\} \quad \text{and} \quad \delta = \min\{d_j \colon j = 1, 2, \ldots n\}.$$

If $[A - D, A]$ is a semipositive interval, then $\delta < \alpha$.

Proof First observe that because A is a semipositive matrix, the quantity α is a well-defined positive number because the maximum is taken over the intersection of the convex cone $K_+(A)$ and the numerical range of A, which is a compact set.

If $[A - D, A]$ is semipositive, then there exists $x > 0$ with $\|x\|_2 = 1$ and $(A - D)x > 0$. That is, $Ax > Dx \geq 0$ and so

$$\alpha \geq x^T Ax > x^T Dx \geq \delta$$

because the numerical range of D is contained in $[\delta, \infty)$. □

Remark 3.7.20 If A is semipositive, the quantity α in Theorem 3.7.19 is positive, which implies that the numerical range of A intersects the positive real line and so $\lambda_{max}\left(\frac{A+A^T}{2}\right) > 0$. The converse is not true, even if one assumes that $A + A^T$ is a nonnegative matrix, as can be seen by the non-semipositive matrix $A = \left[\begin{smallmatrix} 1 & 1 \\ -1 & 0 \end{smallmatrix}\right]$.

Theorem 3.7.21 Let $A \in M_n(\mathbb{R})$ be semipositive and let $u, v \in \mathbb{R}^n_+$. If $[A - uv^T, A]$ is a semipositive interval, then there exists $x \in K_+(A)$ such that

$$\left(u^T x\right)\left(v^T x\right) < x^T Ax.$$

Proof If $[A - uv^T, A]$ is semipositive, then there exists $x > 0$ such that $\left(A - uv^T\right)x > 0$. That is, $Ax > \left(v^T x\right)u \geq 0$, and so $x \in K_+(A)$ with

$$x^T Ax > \left(v^T x\right)\left(u^T x\right).$$ □

Theorem 3.7.22 Let $A \in M_n(\mathbb{R})$ be semipositive and let e denote the all-ones vector in $\mathbb{R}^n$.

(a) Let $A^\dagger \geq 0$, $u, v \in \mathbb{R}^n_+$, $u \in R(A)$ such that $v^T A^\dagger u < 1$. Then $[A - uv^T, A]$ is a semipositive interval.
(b) Let $A^\dagger \geq 0$ and set $B = A^\dagger = [b_{ij}]$. Suppose that $\sum_{i,j=1}^n b_{ij} < 1$ and $e \in R(A)$. Then $[A - ee^T, A]$ is a semipositive interval.
(c) Let A be minimally semipositive and set $C = A^{-1} = [c_{ij}]$. Suppose that $\sum_{i,j=1}^n c_{ij} < 1$. Then $A - ee^T$ is a minimally semipositive matrix so that $[A - ee^T, A]$ is a minimal semipositive interval.

Proof (a) Let $x = A^\dagger u \geq 0$. Note that because $u \in R(A)$, we have $AA^\dagger u = u$. Then

$$\begin{aligned}(A - uv^T)x &= AA^\dagger u - \left(v^T A^\dagger u\right)u \\ &= u - \left(v^T A^\dagger u\right)u \\ &= \left(1 - v^T A^\dagger u\right)u \geq 0,\end{aligned}$$

i.e., $A - uv^T$ is semipositive. It follows that $[A - uv^T, A]$ is a semipositive interval.

(b) Set $u = v = e$. Then $u, v \geq 0,\ u \in R(A)$ and

$$1 - v^T A^{\dagger} u = 1 - e^T A^{\dagger} e = 1 - \sum_{i,j=1}^{n} b_{ij} > 0.$$

The proof now follows from part (a).

(c) Because A is minimally semipositive, it follows that A^{-1} exists and $A^{-1} \geq 0$. By the Sherman–Woodbury formula (e.g., see [HJ91, p. 19]) for the inverse of a rank-one perturbation, we have

$$\left(A - ee^T\right)^{-1} = A^{-1} + \frac{1}{\mu} A^{-1} ee^T A^{-1},$$

where $\mu = 1 - e^T A^{-1} e > 0$. Clearly, $\left(A - ee^T\right)^{-1} \geq 0$ (so that $A - ee^T$ is also minimally semipositive). By Theorem 3.7.17, $[A - ee^T, A]$ is inverse positive. In particular, every such matrix is minimally semipositive, completing the proof. □

3.8 Strictly Semimonotone Matrices

Semimonotone matrices $A \in M_n(\mathbb{R})$ are defined as those matrices for which the operation Ax does not negate all the positive entries of any nonzero, entry-wise nonnegative vector x. If Ax preserves at least one positive entry for every such x, we refer to A as strictly semimonotone. In that respect, strictly semimonotone matrices generalize the class of P-matrices to be studied in Chapter 4, which preserve the sign of a nonzero entry of every nonzero $x \in \mathbb{R}^n$. In fact, it follows that every P-matrix is strictly semimonotone. More important, strictly semimonotone matrices are intimately related to matrix semipositivity, prompting us to review their properties in this section. For instance, as we will see below, A is strictly semimonotone if and only if A itself and all of its proper principal submatrices are semipositive.

We continue with a formal discussion of strictly semimonotone matrices. Additional results, history, and details can be found in [CPS92] and [TW19].

Definition 3.8.1 A matrix $A \in M_n(\mathbb{R})$ is **strictly semimonotone** if for each nonzero, nonnegative $x \in \mathbb{R}^n$, there exists $k \in \{1, 2, \ldots, n\}$ such that $x_k > 0$ and $(Ax)_k > 0$.

The following simple observations can be made immediately. First, if e_k is the k-th column of the $n \times n$ identity matrix and $A \in M_n(\mathbb{R})$ is strictly semimonotone, then the k-th entry of Ae_k must be positive, that is, every strictly semimonotone matrix must have positive diagonal entries. Moreover, by letting x be any vector whose entries indexed by some index set $\alpha \subseteq \{1, 2, \ldots, n\}$ are positive and whose other entries are zero, we can see that $(Ax)_k > 0$ for some $k \in \alpha$. Thus, $A[\alpha]x[\alpha]$ has a positive entry, and so $A[\alpha]$ is strictly semimonotone. Hence, every principal submatrix of a strictly semimonotone matrix must be semimonotone. It is then easy to see that the following theorem holds.

Theorem 3.8.2 A matrix $A \in M_n(\mathbb{R})$ is strictly semimonotone if and only if every proper principal submatrix of A is strictly semimonotone, and for every $x > 0$, there exists $k \in \langle n \rangle$ such that $(Ax)_k > 0$.

Some important facts about strictly semimonotone matrices are summarized in the next theorem.

Theorem 3.8.3

(a) Every square positive matrix is strictly semimonotone.
(b) Every P-matrix is strictly semimonotone.
(c) All strictly copositive matrices are strictly semimonotone.
(d) $A \in M_n(\mathbb{R})$ is strictly semimonotone if and only if A and all its proper principal submatrices are semipositive.
(e) $A \in M_n(\mathbb{R})$ is strictly semimonotone if and only if A^T is strictly semimonotone.
(f) $A \in M_n(\mathbb{R})$ is strictly semimonotone if and only if the Linear Complementarity Problem, LCP(q, A) (see Section 4.9.1) has a unique solution for every $q \geq 0$.

Proof To show (a), let A be a positive matrix. For any $0 \neq x \geq 0$, if $x_k > 0$, then because A does not contain any nonpositive entries, $(Ax)_k > 0$. Next, we show (b). If A is a P-matrix, then for each $x \neq 0$, there exists a k such that $x_k(Ax)_k > 0$ (see Theorem 4.3.4). Thus, because this holds for every $0 \neq x \geq 0$, A is strictly semimonotone. To show (c), suppose A is a strictly copositive, then for any $0 \neq x \geq 0$, we have that $x^T Ax > 0$. Thus, there must be an index k such that $x_k > 0$ and $(Ax)_k > 0$, and so A is strictly semimonotone. For a proofs of clauses (d)–(f), see [CPS92]. □

The proofs of the following result are straightforward.

Theorem 3.8.4 Let $A \in M_n(\mathbb{R})$ be a strictly semimonotone matrix. If $B \in M_n(\mathbb{R})$ is a nonnegative matrix, then $A + B$ is strictly semimonotone.

Theorem 3.8.5 Suppose that all proper principal submatrices of $A \in M_n(\mathbb{R})$ are strictly semimonotone. If A has a row of nonnegative entries or a column of nonnegative entries, then A is strictly semimonotone.

Proof We will show that if $A \in M_n(\mathbb{R})$ has all proper principal submatrices strictly semimonotone and if it has a column of nonnegative entries, then A is strictly semimonotone. The case when A has a row of nonnegative entries would then follow from the Theorem 3.8.3. Let $a_1, a_2, \ldots, a_n \in \mathbb{R}^n$ be the columns of A and suppose, without loss of generality, that $a_1 \geq 0$. Suppose A is not strictly semimonotone. Then, because all proper principal submatrices arestrictly semimonotone, there exists an $x = [x_1\ x_2\ \ldots,\ x_n]^T > 0$ such that $Ax \leq 0$. Let $y = [0\ x_2\ x_3\ \ldots,\ x_n]^T$ and note that there must exist $k \in \{2, 3, \ldots, n\}$ such that $(Ay)_k > 0$, because the principal submatrix $A[\{2, 3, \ldots, n\}]$ is strictly semimonotone. However, because $Ax \leq 0$ and $x_1 a_1 \geq 0$, we have

$$
\begin{aligned}
Ay = \begin{bmatrix} a_1 & a_2 & a_3 & \cdots & a_n \end{bmatrix} \begin{bmatrix} 0 \\ x_2 \\ x_3 \\ \vdots \\ x_n \end{bmatrix} \\
= 0a_1 + x_2 a_2 + x_3 a_3 + \cdots + x_n a_n \\
= Ax - x_1 a_1 \leq 0,
\end{aligned}
$$

a contradiction. □

The following two results are simple consequences of the definition of strict semimonotonicity.

Theorem 3.8.6 Let $P \in M_n(\mathbb{R})$ be a permutation matrix. Then $A \in M_n(\mathbb{R})$ is strictly semimonotone if and only if PAP^T is strictly semimonotone.

Theorem 3.8.7 Let $A \in M_n(\mathbb{R})$ and let $D = \text{diag}(d_1, d_2, \ldots, d_n)$ be a diagonal matrix with $d_i > 0$. Then the following statements are equivalent.

(a) A is strictly semimonotone.
(b) DA is strictly semimonotone.
(c) AD is strictly semimonotone.

Theorem 3.8.8 Let $A \in M_n(\mathbb{R})$ be a block upper triangular matrix with square diagonal blocks. Then A is strictly semimonotone if and only if each diagonal block is strictly semimonotone.

Proof

($\Leftarrow$) Suppose A is a strictly semimonotone block upper triangular matrix with square diagonal blocks. Each such block is a principal submatrix of A and so, by Theorem 3.8.2, it is also strictly semimonotone.

($\Rightarrow$) Let A be a block upper triangular matrix with square diagonal blocks that are strictly semimonotone. We will proceed by induction on the number of diagonal blocks. The base case is when A is a two-block upper triangular matrix with both diagonal blocks strictly semimonotone; i.e., $A = \begin{bmatrix} A_1 & B \\ 0 & A_2 \end{bmatrix} \in M_n(\mathbb{R})$, where $A_1 \in M_r(\mathbb{R})$ and $A_2 \in M_{n-r}(\mathbb{R})$ are strictly semimonotone. Suppose A is not strictly semimonotone. Take a set $\alpha \subseteq \{1, 2, \ldots, n\}$ of minimum cardinality for which the principal submatrix $A[\alpha]$ is not strictly semimonotone. Note that we must have $\alpha = \alpha_1 \cup \alpha_2$ where $\alpha_1 \subseteq \{1, 2, \ldots, r\}$ is nonempty and $\alpha_2 \subseteq \{r+1, r+2, \ldots, n\}$ is nonempty; otherwise, the principal submatrix $A[\alpha]$ would be strictly semimonotone. Then,

$$A[\alpha] = \begin{bmatrix} A[\alpha_1] & \tilde{B} \\ 0 & A[\alpha_2] \end{bmatrix},$$

which is also a two-block upper triangular matrix with the diagonal blocks strictly semimonotone. Now, the minimum cardinality of α implies that there must be an $x > 0$ such that $A[\alpha]x \leq 0$. However, this would imply that $A[\alpha_2]z \leq 0$ for some positive vector z, a contradiction because $A[\alpha_2]$ is strictly semimonotone. Hence, A must be strictly semimonotone and the base case holds.

Now suppose that every block upper triangular matrix with k diagonal blocks all of which are strictly semimonotone is strictly semimonotone. Consider a matrix A with $k+1$ diagonal blocks. We know that the submatrix of A containing the first k diagonal blocks is strictly semimonotone by the inductive hypothesis, so we can consider this submatrix to be a single diagonal block. Then A consists of two diagonal blocks, which are each strictly semimonotone. Thus, A is strictly semimonotone. □

Theorem 3.8.9 Suppose $A \in M_n(\mathbb{R})$ with all proper principal submatrices being strictly semimonotone. Then A is strictly semimonotone if and only if for all diagonal matrices $D \in M_n(\mathbb{R})$ with nonnegative diagonal entries, $A + D$ does not have a positive nullvector.

Proof

($\Rightarrow$) We will prove the contrapositive. Suppose there exists a diagonal matrix D with nonnegative diagonal entries such that $A + D$ has a positive null vector.

Then there exists an $x > 0$ such that $(A + D)x = 0$. So $Ax = -Dx \leq 0$. Thus, A is not strictly semimonotone.

($\Leftarrow$) Suppose A is not strictly semimonotone. We want to show that there exists a diagonal matrix D with nonnegative diagonal entries such that $A+D$ has a positive null vector. Because A is not strictly semimonotone (but all proper principal submatrices are), there exists an $x > 0$ such that $Ax \leq 0$. Now, let $D = [d_{ij}]$ be a diagonal matrix where

$$d_{ii} = -\frac{(Ax)_i}{x_i}.$$

Note D has nonnegative diagonal entries and $Dx = -Ax$. Thus, $(A + D)x = 0$. Because $x > 0$, $A + D$ has a positive null vector. □

Let $[0, I]$ denote all $n \times n$ diagonal matrices whose diagonal entries are in $[0, 1]$.

Theorem 3.8.10 A matrix $A \in M_n(\mathbb{R})$ is strictly semimonotone if and only if for all $T \in [0, I]$, $T + (I - T)A$ has no null vector x such that $0 \neq x \geq 0$.

Proof

($\Rightarrow$) The proof is by contradiction. Suppose that A is strictly semimonotone and suppose that for some $T \in [0, I]$ there exists a $0 \neq x \geq 0$ such that $(T + (I - T)A)x = 0$. Let $y = Ax$. Because A is strictly semimonotone, there exists a k such that $x_k > 0$ and $y_k = (Ax)_k > 0$. Notice that

$$\begin{aligned}(T + (I - T)A)x = 0 \quad &\Rightarrow \quad t_{kk}x_k + (1 - t_{kk})(Ax)_k = 0 \\ &\Rightarrow \quad t_{kk}(x_k - y_k) = -y_k.\end{aligned}$$

Now, if $x_k = y_k$, then $0 = y_k$, a contradiction because $y_k > 0$. Thus, $x_k - y_k \neq 0$ and so

$$t_{kk} = -\frac{y_k}{x_k - y_k}.$$

Now suppose $x_k - y_k > 0$. In this case, because $t_{kk} \geq 0$, we get that

$$-\frac{y_k}{x_k - y_k} \geq 0 \quad \Rightarrow \quad -y_k \geq 0 \quad \Rightarrow \quad y_k \leq 0,$$

a contradiction because $y_k > 0$. Finally, suppose that $x_k - y_k < 0$. In this case, because $t_{kk} \leq 1$, we get that

$$-\frac{y_k}{x_k - y_k} \leq 1 \quad \Rightarrow \quad -y_k \geq x_k - y_k \quad \Rightarrow \quad 0 \geq x_k,$$

a contradiction because $x_k > 0$. Thus, no such null vector can exist.

($\Leftarrow$) We will prove the contrapositive. Suppose that A is not strictly semimonotone. Then there exists a $0 \neq x \geq 0$ such that $x_k(Ax)_k \leq 0$ for each k. Let $y = Ax$. Note, if $x_k > 0$, we must have $y_k = 0$ or $y_k < 0$. If $x_k = 0$, then we can have $y_k = 0, y_k < 0$, or $y_k > 0$. Now, we will find a $T \in [0, I]$ such that $(T + (I - T)A)x = 0$. Let

$$t_{ii} = \begin{cases} 0 & \text{if } x_i > 0 \text{ and } y_i = 0 \\ -\dfrac{y_i}{x_i - y_i} & \text{if } x_i > 0 \text{ and } y_i < 0 \\ 0 & \text{if } x_i = 0 = y_i \\ 1 & \text{if } x_i = 0 \text{ and } y_i > 0 \\ 1 & \text{if } x_i = 0 \text{ and } y_i < 0. \end{cases}$$

Notice that if $x_i > 0$ and $y_i < 0$, then $0 \leq -\frac{y_i}{x_i - y_i} \leq 1$. Suppose not. If $-\frac{y_i}{x_i - y_i} < 0$, then because $x_i - y_i > 0$, $-y_i < 0$ and so $y_i \geq 0$, a contradiction. If $-\frac{y_i}{x_i - y_i} > 1$, then $-y_i > x_i - y_i$, and so $x_i < 0$, a contradiction. Hence, we see that $0 \leq t_{ii} \leq 1$. Also, notice that

$$((I - T)Ax)_i = (1 - t_{ii})y_i = -t_{ii}x_i = -(Tx)_i.$$

Hence, $(T + (I - T)A)x = 0$. □

For proofs of the following results, we refer the reader to [TW19].

Theorem 3.8.11 Given the spectrum σ of any real $n \times n$ matrix with positive trace, there exists a strictly semimonotone matrix $A \in M_n(\mathbb{R})$ such that $\sigma(A) = \sigma$.

A **signature matrix** $S \in M_n(\mathbb{R})$ is a diagonal matrix whose diagonal entries belong to $\{-1, 1\}$.

Theorem 3.8.12 [cf. Theorem 4.3.8] The following are equivalent for $A \in M_n(\mathbb{R})$:

(a) A is a P-matrix.
(b) SAS is semipositive for all signature matrices S.
(c) SAS is strictly semimonotone for all signature matrices S.

In [TW19], the term **almost (strictly) semimonotone** $A \in M_n(\mathbb{R})$ is introduced when all proper principal submatrices of A are (strictly) semimonotone and there exists an $x > 0$ such that $Ax < 0$ ($Ax \leq 0$). Then the following results are shown.

Theorem 3.8.13 If $A \in M_n(\mathbb{R})$ is almost semimonotone, then $-A$ is **MSP**

Theorem 3.8.14 Suppose $A \in M_n(\mathbb{R})$ $(n > 1)$ is almost strictly semimonotone and not semimonotone. Then A^{-1} exists and $A^{-1} < 0$.

4

P-Matrices

4.1 Introduction

Recall that $A \in M_n(\mathbb{C})$ is called a **P-matrix** ($A \in \mathbf{P}$), if all its principal minors are positive, i.e.,

$$\det A[\alpha] > 0 \quad \text{for all} \quad \alpha \subseteq \langle n \rangle.$$

The P-matrices encompass such notable classes as the positive definite matrices (**PD**), the (inverse) M-matrices (**M**, **IM**), the totally positive matrices (**TP**), as well as the H^{+}-matrices. As we will see, P-matrices are semipositive (**SP**). The study of P-matrices originated in the context of these classes in the work of Ostrowski, Fan, Koteljanskii, Gantmacher and Krein, Taussky, Fiedler and Pták, Tucker, as well as Gale and Nikaido. Some classical references to this early work include [FP62, FP66, GN65, GK1935, Tau58].

The first systematic study of P-matrices appears in the work of Fiedler and Pták [FP66]. Since then, the class **P** and its subclasses have proven a fruitful research subject, judged by the attention received in the matrix theory community and the continuing interest generated by the applications of P-matrices in the mathematical sciences. P-matrices play an important role in a wide range of applications, including the linear complementarity problem, global univalence of maps, linear differential inclusion problems, interval matrices, and computational complexity. Some of these applications are discussed in this chapter; see also [BEFB94, CPS92, BP94, Parth96].

Of particular concern is the ability to decide as efficiently as possible whether an n-by-n matrix is in **P** or not, referred to as the **P-problem**. It has received attention largely due to its inherent computational complexity. Motivated by the P-problem and questions about the spectra of P-matrices, in this chapter we address the need to construct (generic and special) P-matrices for purposes of experimentation, as well as theoretical and algorithmic development. To this end, in this chapter we provide a review of (i) basic properties of P-matrices and operations that preserve them, (ii) techniques to generate special and generic P-matrices, (iii) numerical methods to detect P-matrices, and (iv) manifestations of P-matrices in various mathematical contexts. This approach affords us the opportunity to review well-known results, some of them presented under new light, as well as to bring forth some less known and some newer results on P-matrices. Other comprehensive treatments of P-matrix theory can be found in [BP94, Fie86, HJ91].

This chapter unfolds as follows: Section 4.2 contains preliminary material specific to this chapter. Section 4.3 reviews basic properties and characterizations of P-matrices, including mappings of **P** into itself. Section 4.4 contains facts and questions about the eigenvalues of P-matrices. Section 4.5 discusses P-matrices with additional properties, including a closer examination of a special subclass of the P-matrices (mimes) that encompasses the M-matrices and their inverses. Section 4.6 provides an algorithmic resolution of the general P-problem, as well as approaches suitable for the detection of special subclasses of the P-matrices. Section 4.7 combines results from previous sections to provide a method that can generate every P-matrix. Section 4.8 concerns the topological closure of **P**, specifically, the adaptation of results for **P** to the class of matrices with nonnegative principal minors. Section 4.9 reviews manifestations and applications of P-matrices in various mathematical contexts. The chapter concludes with Section 4.10 in which further P-matrix considerations, generalizations, and related facts are collected, as well as comments on work not covered in this book.

4.2 Notation, Definitions, and Preliminaries

4.2.1 Matrix Transforms

Definition 4.2.1 Given a nonempty $\alpha \subseteq \langle n \rangle$ and provided that $A[\alpha]$ is invertible, we define the **principal pivot transform** of $A \in M_n(\mathbb{C})$ relative to α as the matrix $\text{ppt}\,(A,\alpha)$ obtained from A by replacing

$$A[\alpha] \quad \text{by } A[\alpha]^{-1}, \qquad A[\alpha,\overline{\alpha}] \text{ by } -A[\alpha]^{-1}A[\alpha,\overline{\alpha}],$$
$$A[\overline{\alpha},\alpha] \text{ by } A[\overline{\alpha},\alpha]A[\alpha]^{-1} \quad \text{and} \quad A[\overline{\alpha}] \text{ by } A/A[\alpha].$$

By convention, $\text{ppt}\,(A,\emptyset) = A$. To illustrate this definition, when $\alpha = \{1,2,\ldots,k\}$ $(0 < k < n)$, we have that

$$\text{ppt}\,(A,\alpha) = \begin{bmatrix} A[\alpha]^{-1} & -A[\alpha]^{-1}A[\alpha,\overline{\alpha}] \\ A[\overline{\alpha},\alpha]A[\alpha]^{-1} & A/A[\alpha] \end{bmatrix}.$$

The effect of applying a principal pivot transform is as follows. Suppose that $A \in M_n(\mathbb{C})$ is partitioned in blocks as

$$A = \begin{bmatrix} A_{11} & A_{12} \\ A_{21} & A_{22} \end{bmatrix} \tag{4.2.1}$$

and further suppose that A_{11} is invertible. Consider then the principal pivot transform relative to the leading block

$$B = \begin{bmatrix} (A_{11})^{-1} & -(A_{11})^{-1}A_{12} \\ A_{21}(A_{11})^{-1} & A_{22} - A_{21}(A_{11})^{-1}A_{12} \end{bmatrix}. \tag{4.2.2}$$

The matrices A and B are related as follows: If $x = \begin{bmatrix} x_1^T \; x_2^T \end{bmatrix}^T$ and $y = \begin{bmatrix} y_1^T \; y_2^T \end{bmatrix}^T$ in $\mathbb{C}^n$ are partitioned conformally to A, then

$$A\begin{bmatrix} x_1 \\ x_2 \end{bmatrix} = \begin{bmatrix} y_1 \\ y_2 \end{bmatrix} \text{ if and only if } B\begin{bmatrix} y_1 \\ x_2 \end{bmatrix} = \begin{bmatrix} x_1 \\ y_2 \end{bmatrix}.$$

For a review of the properties and applications of the principal pivot transform, see [Tsa00].

Definition 4.2.2 For $A \in M_n(\mathbb{C})$ with $-1 \notin \sigma(A)$, the fractional linear map

$$F_A = (I + A)^{-1}(I - A)$$

is called the **Cayley transform** of A. The map $A \to F_A$ is an involution, namely,

$$A = (I + F_A)^{-1}(I - F_A).$$

4.2.2 More Matrix Classes

Below we recall and introduce some matrix classes referenced in this chapter.

- We let $\mathbf{P_M}$ denote the class of matrices all of whose positive integer powers are in $\mathbf{P}$.
- A **positive stable** matrix $A \in M_n(\mathbb{C})$ is a matrix all of whose eigenvalues lie in the *open* right-half plane.
- $A = [a_{ij}] \in M_n(\mathbb{C})$ is **row diagonally dominant** if for all $i \in \langle n \rangle$,

$$|a_{ii}| > \sum_{j \neq i} |a_{ij}|.$$

Note that in our terminology the diagonal dominance is strict. Due to Gershgorin's Theorem (see Chapter 1), row diagonally dominant matrices with positive diagonal entries are positive stable.

- We call $A = [a_{ij}] \in M_n(\mathbb{R})$ a **B-matrix** if for each $i \in \langle n\rangle$ and all $j \in \langle n\rangle \setminus \{i\}$,

$$\sum_{k=1}^{n} a_{ik} > 0 \quad \text{and} \quad \frac{1}{n} \sum_{k=1}^{n} a_{ik} > a_{ij};$$

namely, the row sums of a B-matrix are positive, and the row averages dominate the off-diagonal entries. The properties and applications of B-matrices are studied in [Pen01].
- A **Z-matrix (Z)** is a square matrix all of whose off-diagonal entries are nonpositive. An invertible **M-matrix (M)** is a positive stable Z-matrix or, equivalently, a semipositive Z-matrix. An **inverse M-matrix (IM)** is the inverse of an M-matrix. An **MMA-matrix** is a matrix all of whose positive integer powers are irreducible M-matrices (see Section 4.2.3 for the definition of irreducibility).

  An M-matrix A can be written as $A = sI - B$, where $B \geq 0$ and $s \geq \rho(B)$. The Perron–Frobenius Theorem applied to B and B^T implies that A possesses right and left nonnegative eigenvectors x, y, respectively, corresponding to the eigenvalue $s - \rho(B)$. We refer to x and y as the **(right) Perron eigenvector** and the **left Perron eigenvector** of A, respectively. When B is also irreducible, $s - \rho(B)$ is a simple eigenvalue of A, and we may take $x > 0$ and $y > 0$.
- The **comparison matrix** of $A = [a_{ij}] \in M_n(\mathbb{C})$, denoted by $\mathcal{M}(A) = [b_{ij}]$, is defined by

$$b_{ij} = \begin{cases} -|a_{ij}| & \text{if } i \neq j \\ |a_{ii}| & \text{otherwise.} \end{cases}$$

- We call $A \in M_n(\mathbb{C})$ an **H-matrix** if $\mathcal{M}(A)$ is an M-matrix.
- The following class of matrices was defined by Pang in [Pan79a, Pan79b], extending notions introduced in the work of Mangasarian and Dantzig. Such matrices were called **hidden Minkowski matrices** in [Pan79b]; we adopt a different name for them, indicative of their matricial nature and origin.

Definition 4.2.3 Consider a matrix $A \in M_n(\mathbb{R})$ of the form

$$A = (s_1 I - P_1)\,(s_2 I - P_2)^{-1},$$

where $s_1,\ s_2 \in \mathbb{R}$, $P_1,\ P_2 \geq 0$, such that for some vector $u \geq 0$,

$$P_1 u < s_1 u \ \text{ and } \ P_2 u < s_2 u.$$

We call A a **mime**, which is an acronym for **M**-matrix and **I**nverse **M**-matrix **E**xtension, because the class of mimes contains the M-matrices (by taking

$P_2 = 0, s_2 = 1$) and their inverses (by taking $P_1 = 0, s_1 = 1$). We refer to the nonnegative vector u above as a **common semipositivity vector** of A.

- We conclude this section with the definition of a type of orthogonal matrix to be used in generating matrices all of whose powers are in **P**. An n-by-n **Soules matrix** R is an orthogonal matrix with columns $\{w_1, w_2, \ldots, w_n\}$ such that $w_1 > 0$ and $R\Lambda R^T \geq 0$ for every

$$\Lambda = \text{diag}(\lambda_1, \lambda_2, \ldots, \lambda_n)$$

with $\lambda_1 \geq \lambda_2 \geq \cdots \geq \lambda_n \geq 0$. Soules matrices can be constructed starting with an arbitrary positive vector w_1 such that $\|w_1\|_2 = 1$; for details see [ENN98, Sou83].

4.2.3 Sign Patterns and Directed Graphs

We call a diagonal matrix S whose diagonal entries belong to $\{-1, 1\}$ a **signature matrix**. Note that $S = S^{-1}$; thus we refer to SAS as a **signature similarity** of A.

Matrix $A \in M_n(\mathbb{R})$ is called **sign nonsingular** if every matrix with the same sign pattern as A (see Section 3.3) is nonsingular.

The **directed graph**, $D(A)$, of $A = [a_{ij}] \in M_n(\mathbb{C})$ consists of the set of vertices $\{1, \ldots, n\}$ and the set of directed edges (i,j) connecting vertex i to vertex j if and only if $a_{ij} \neq 0$. We say $D(A)$ is **strongly connected** if any two distinct vertices i,j are connected by a path of edges $(i, i_1), (i_1, i_2), \ldots, (i_{k-1}, i_k), (i_k, j)$. When $D(A)$ is strongly connected, we refer to A as an **irreducible matrix**. A **cycle of length** k in $D(A)$ consists of edges

$$(i_1, i_2), (i_2, i_3), \ldots, (i_{k-1}, i_k), (i_k, i_1),$$

where the vertices $i_1, i_2, \ldots, i_k$ are distinct. The nonzero diagonal entries of A correspond to cycles of length 1 in $D(A)$. The **signed directed graph** of $A \in M_n(\mathbb{R})$, $S(A)$, is obtained from $D(A)$ by labeling each edge (i,j) with the sign of a_{ij}. We define the **sign of a cycle** on the vertices $\{i_1, i_2, \ldots, i_k\}$ as above to be the sign of the product $a_{i_1 i_2} a_{i_2 i_3} \cdots a_{i_{k-1} i_k} a_{i_k i_1}$.

We denote by M_n^k ($k \leq n$) the set of matrices $A \in M_n(\mathbb{R})$ with nonzero diagonal entries such that the length of the longest cycle in $D(A)$ is no more than k. For matrices in M_n^k, we adopt the following notation: $A \in \mathbf{P}_n^k$ if $A \in M_n^k$ is a P-matrix; $A \in S_n^k$ if all the cycles in $S(-A)$ are signed negatively.

4.3 Basic Properties of P-Matrices

A basic observation comes first.

Observation 4.3.1 A block triangular matrix with square diagonal blocks is a P-matrix if and only if each diagonal block is a P-matrix. In particular, the direct sum of P-matrices is a P-matrix.

Proof The result follows from the following two facts. The determinant of a block triangular matrix with square diagonal blocks is the product of the determinants of the diagonal blocks. Also, if A is a block triangular matrix with square diagonal blocks, all principal submatrices of A are block triangular matrices whose diagonal blocks are principal submatrices of A. □

We proceed with a review of transformations that map **P** *into* itself.

Theorem 4.3.2 Let $A \in M_n(\mathbb{C})$ be a P-matrix ($A \in \mathbf{P}$). Then the following hold.

(1) $A^T \in \mathbf{P}$.
(2) $QAQ^T \in \mathbf{P}$ for every permutation matrix Q.
(3) $SAS \in \mathbf{P}$ for every signature matrix S.
(4) $DAE \in \mathbf{P}$ for all diagonal matrices D, E such that DE has positive diagonal entries.
(5) $A+D \in \mathbf{P}$ for all diagonal matrices D with nonnegative diagonal entries.
(6) $A[\alpha] \in \mathbf{P}$ for all nonempty $\alpha \subseteq \langle n \rangle$.
(7) $A/A[\alpha] \in \mathbf{P}$ for all $\alpha \subseteq \langle n \rangle$.
(8) $\operatorname{ppt}(A, \alpha) \in \mathbf{P}$ for all $\alpha \subseteq \langle n \rangle$.
In particular, when $a = \langle n \rangle$, we obtain that $\operatorname{ppt}(A, \langle n \rangle) = A^{-1} \in \mathbf{P}$.
(9) $I + F_A \in \mathbf{P}$ and $I - F_A \in \mathbf{P}$.
(10) $TI + (I - T)A \in \mathbf{P}$ for all $T \in [0, I]$.

Proof Clauses (1)–(4) and (6) are direct consequences of determinantal properties and the definition of a P-matrix.

(5) Notice that if $A = [a_{ij}] \in \mathbf{P}$, then $\frac{\partial \det A}{\partial a_{ii}} = \det A[\overline{\{i\}}] > 0$, that is, $\det A$ is an increasing function of the diagonal entries. Thus, as the diagonal entries of D are added in succession to the diagonal entries of A, the determinant of A and, similarly, all principal minors of A remain positive.

(7) Because $A \in \mathbf{P}$, $A[\alpha]$ is invertible and A^{-1}, up to a permutation similarity, has the block representation (see, e.g., [HJ13, (0.7.3) p. 18])

$$\begin{bmatrix} (A/A[\overline{\alpha}])^{-1} & -A[\alpha]^{-1}A[\alpha,\overline{\alpha}](A/A[\alpha])^{-1} \\ -(A/A[\alpha])^{-1}A[\overline{\alpha},\alpha]A[\alpha]^{-1} & (A/A[\alpha])^{-1} \end{bmatrix}.$$

Therefore, every principal submatrix of A^{-1} is of the form $(A/A[\alpha])^{-1}$ for some $\alpha \subseteq \langle n \rangle$ and has determinant $\det A[\alpha]/\det A > 0$. This shows that $A^{-1} \in \mathbf{P}$ and, in turn, that $(A/A[\alpha])^{-1}$ and thus $A/A[\alpha]$ are P-matrices for every $\alpha \subseteq \langle n \rangle$.

(8) Let $A \in \mathbf{P}$ and consider first the case where α is a singleton; without loss of generality assume that $\alpha = \{1\}$. Let $B = \text{ppt}\,(A, \alpha) = [b_{ij}]$. By definition, the principal submatrices of B that do not include entries from the first row coincide with the principal submatrices of $A/A[\alpha]$ and thus, by (7), have positive determinants. The principal submatrices of B that include entries from the first row of B are equal to the corresponding principal submatrices of the matrix B' obtained from B using $b_{11} = (A[\alpha])^{-1} > 0$ as the pivot and eliminating the nonzero entries below it. Notice that

$$B' = \begin{bmatrix} 1 & 0 \\ -A[\overline{\alpha}, \alpha] & I \end{bmatrix} \begin{bmatrix} b_{11} & -b_{11}A[\alpha, \overline{\alpha}] \\ A[\overline{\alpha}, \alpha]b_{11} & A/A[\alpha] \end{bmatrix} = \begin{bmatrix} b_{11} & -b_{11}A[\alpha, \overline{\alpha}] \\ 0 & A[\overline{\alpha}] \end{bmatrix}.$$

That is, B' is itself a P-matrix for it is block upper triangular, and the diagonal blocks are P-matrices. It follows that all the principal minors of B are positive and thus $B \in \mathbf{P}$. Next, consider the case $\alpha = \{i_1, i_2, \ldots, i_k\} \subseteq \langle n \rangle$ with $k \geq 1$. By our arguments so far, the sequence of matrices

$$A_0 = A, \;\; A_j = \text{ppt}\,(A_{j-1}, \{i_j\}),\; j = 1, 2, \ldots, k$$

is well defined and comprises P-matrices. Moreover, from the uniqueness of $B = \text{ppt}\,(A, \alpha)$ shown in [Tsa00, theorem 3.1], it follows that $A_k = \text{ppt}\,(A, \alpha) = B$ and thus $B \in \mathbf{P}$.

(9) $A \in \mathbf{P}$ is nonsingular and $-1 \notin \sigma(A)$; otherwise, (5) would be violated for $D = I$. Hence F_A is well defined. It can also be verified that

$$I + F_A = 2(I + A)^{-1} \quad \text{and} \quad I - F_A = 2(I + A^{-1})^{-1}.$$

As addition of positive diagonal matrices and inversion are operations that preserve P-matrices (see (5) and (8)), it follows that $I - F_A$ and $I + F_A$ are P-matrices.

(10) Let $A \in \mathbf{P}$ and $T = \text{diag}(t_1, \ldots, t_n) \in [0, I]$. Because T and $I - T$ are diagonal,

$$\det(TI + (I - T)A) = \sum_{\alpha \subseteq \langle n \rangle} \prod_{i \notin \alpha} t_i \, \det\left(((I - T)A)\,[\alpha]\right).$$

As $t_i \in [0, 1]$ and $A \in \mathbf{P}$, all summands in this determinantal expansion are nonnegative. Unless $T = 0$, in which case $TI + (I - T)A = A$, one of these summands is positive. Hence $\det(A) > 0$. The exact same argument can be applied to any principal submatrix of A, proving that $A \in \mathbf{P}$. $\square$

Remarks 4.3.3 The following comments pertain to Theorem 4.3.2.

(i) It is straightforward to argue that each one of the clauses (1)–(5), (7), and (10) represents a necessary and sufficient condition that A be a P-matrix.

(ii) By repeated application of (8), it follows that any one principal pivot transform of A is a P-matrix if and only if every principal pivot transform of A is a P-matrix; see [Tsa00].

(iii) A more extensive analysis of the relations among the Cayley transforms of P-matrices and other matrix positivity classes is undertaken in [FT02].

We continue with more basic facts and characterizations of P-matrices. First is the classical characterization of real P-matrices as those matrices that do not "reverse the sign of any nonzero real vector."

Theorem 4.3.4 If $A \in M_n(\mathbb{R})$, then $A \in \mathbf{P}$ if and only if for each nonzero $x \in \mathbb{R}^n$, there exists $j \in \langle n \rangle$ such that $x_j(Ax)_j > 0$.

Proof Suppose that $A \in M_n(\mathbb{R})$ and that there exists $x \in \mathbb{R}^n$ such that for all $j \in \langle n \rangle$, $x_j(Ax)_j \leq 0$. Then there exists positive diagonal matrix D such that $Ax = -Dx$, i.e., $(A + D)x = 0$. By Theorem 4.3.2 (5), A is not a P-matrix. This proves that when $A \in M_n(\mathbb{C})$ is a P-matrix, then for every nonzero $x \in \mathbb{R}^n$, there exists $j \in \langle n \rangle$ such that $x_j(Ax)_j > 0$.

Suppose now that $A \in M_n(\mathbb{R})$ and that for each nonzero $x \in \mathbb{R}^n$, there exists $j \in \langle n \rangle$ such that $x_j(Ax)_j > 0$. Notice that the same holds for every principal submatrix $A[\alpha]$ of A, by simply considering vectors x such that $x[\overline{\alpha}] = 0$. Thus all the real eigenvalues of $A[\alpha]$ are positive, for all nonempty $\alpha \subseteq \langle n \rangle$. As complex eigenvalues come in conjugate pairs, it follows that all the principal minors of A are positive. □

Theorem 4.3.5 Let $A \in M_n(\mathbb{R})$. Then the following are equivalent.

(i) $A \in \mathbf{P}$.

(ii) $Ax \neq 0$ for any $x \in \mathbb{R}^n$ with no zero entries.

(iii) For each nonzero $x \in \mathbb{R}^n$, there exists diagonal matrix D such that $x^T DAx > 0$.

Proof Clearly (i) implies (ii) because A is invertible. To see that (ii) implies (i), if $x \in \mathbb{R}^n$ has no zero entries and $Ax = 0$, then by Theorem 4.3.4, $A \notin \mathbf{P}$. That (i) implies (iii), also follows from Theorem 4.3.4: Let $x \in \mathbb{R}^n$ be nonzero. Then, there exists $j \in \langle n \rangle$ such that $x_j(Ax)_j > 0$. Let $D = \text{diag}(d_1, d_2, \ldots, d_n)$ be a diagonal matrix with $d_k = 1$ for all $k \neq j$. Then we can choose $d_j > 0$ sufficiently large to have $x^T DAx > 0$, showing (i) implies (iii). We can complete the proof by showing that not (ii) implies not (iii). Indeed, if $Ax = 0$ for some $x \in \mathbb{R}^n$ with nonzero entries, then $x^T DAx = 0$ for all diagonal matrices D. □

Theorem 4.3.6 Let $A \in M_n(\mathbb{R})$ be a P-matrix. Then A is a semipositive matrix.

Proof We will prove the contrapositive. By Theorem 1.4.1, if A is not semipositive, then there exists a nonzero $x \geq 0$ such that $A^T x \leq 0$. Then by Theorem 4.3.4, A^T is not a P-matrix and thus A is not a P-matrix. □

Not every semipositive matrix is a P-matrix as shown by the following simple counterexample.

Example 4.3.7 The matrix

$$A = \begin{bmatrix} 1 & 2 \\ 3 & -1 \end{bmatrix}$$

maps the all-ones vector in $\mathbb{R}^2$ to a positive vector, but A is not a P-matrix.

There is, however, a characterization of real P-matrices via semipositivity, first observed in [Al-No88]. It is provided next, along with a simpler proof.

Theorem 4.3.8 $A \in M_n(\mathbb{R})$ is a P-matrix if and only if for every signature matrix $S \in M_n(\mathbb{R})$, SAS is semipositive.

Proof Let $A \in \mathbf{P}$ and $S \in M_n(\mathbb{R})$ be a signature matrix. By Theorem 4.3.2 (3), $SAS \in \mathbf{P}$. If SAS is not semipositive, then by Theorem 1.4.1, there is a nonzero $x \geq 0$ such that $SA^T Sx \leq 0$. Thus $SA^T S \notin \mathbf{P}$ and so $SAS \notin \mathbf{P}$, a contradiction that shows that SAS is semipositive. Conversely, suppose that for every signature matrix $S \in M_n(\mathbb{R})$, SAS is semipositive. By Theorem 1.4.1, for every signature S and every nonzero $x \geq 0$, $SA^T Sx = y \not\leq 0$. Because $A^T(Sx) = Sy$, we have that Sx and Sy have at least one nonzero entry of the same sign. By Theorem 4.3.4, A^T and thus A are P-matrices. □

We conclude this section with a characterization from [JT95] of real P-matrices in factored form. Recall that the matrix interval $[0, I]$ contains all diagonal matrices with entries between 0 and 1, inclusive.

Theorem 4.3.9 Let $A = BC^{-1} \in M_n(\mathbb{R})$. Then $A \in \mathbf{P}$ if and only if the matrix $TB + (I - T)C$ is invertible for every $T \in [0, I]$.

Proof If $A \in \mathbf{P}$, by Theorem 4.3.2 (10), $TI + (I - T)BC^{-1} \in \mathbf{P}$ for every $T \in [0, I]$. Thus $TC + (I - T)B$ is invertible for every $T \in [0, I]$.

For the converse, suppose $TI + (I - T)BC^{-1}$ is invertible for every $T \in [0, I]$ and, by way of contradiction, suppose $BC^{-1} \notin \mathbf{P}$. By Theorem 4.3.4, there exists nonzero $x \in \mathbb{R}^n$ such that $y_j x_j \leq 0$ for every $j \in \langle n \rangle$, where $y = BC^{-1}x$.

Consider then $T = \operatorname{diag}(t_1, t_2, \ldots, t_n)$, where for each $j \in \langle n \rangle$, $t_j \in [0, 1]$ is selected so that $t_j x_j + (1 - t_j) y_j = 0$. It follows that

$$Tx + (I - T)BC^{-1}x = 0,$$

a contradiction that completes the proof. □

Corollary 4.3.10 Let $A \in M_n(\mathbb{R})$. Then $A \in \mathbf{P}$ if and only if the matrix $T + (I - T)A$ is invertible for every $T \in [0, I]$.

Proof The result follows from Theorem 4.3.9 by taking $B = I$ and $C = A^{-1}$. □

4.4 The Eigenvalues of P-Matrices

We begin with some basic observations about the eigenvalues of a P-matrix.

Theorem 4.4.1 Let $A \in M_n(\mathbb{C})$ be a P-matrix with characteristic polynomial $f(t) = \det(tI - A)$. Then the following hold.

(i) $f(t)$ is a monic polynomial with real coefficients that alternate in sign.
(ii) The spectrum of A is closed under complex conjugation, i.e., $\overline{\sigma(A)} = \sigma(A)$.
(iii) For each nonempty $\alpha \subset \langle n \rangle$, every real eigenvalue of $A[\alpha]$ is positive.

Proof
(i) The coefficient of t^k $(k = 0, 1, \ldots, n-1)$ in $f(t)$ is $(-1)^{n-k}$ times the sum of all $k \times k$ principal minors of A. Thus, its sign is $(-1)^{n-k}$.

(ii) By (i), $f(t)$ is a polynomial over the real numbers so its non-real roots come in complex conjugate pairs.

(iii) If for some nonempty $\alpha \subset \langle n \rangle$, $\lambda \leq 0$ were an eigenvalue of $A[\alpha]$, then $A[\alpha] + |\lambda| I$ would be a singular matrix, violating Theorem 4.3.2 (5) applied to the P-matrix $A[\alpha]$. □

The above theorem leads to the following spectral characterization of a P-matrix.

Theorem 4.4.2 Let $A \in M_n(\mathbb{C})$ and its principal submatrices have real characteristic polynomials. Then $A \in \mathbf{P}$ if and only if every real eigenvalue of every principal submatrix of A is positive.

Proof If $A \in \mathbf{P}$ so are its principal submatrices and thus their real eigenvalues are positive by Theorem 4.4.1 (iii). For the converse, let the real eigenvalues of A and its principal submatrices be positive. Then the principal minors of A are

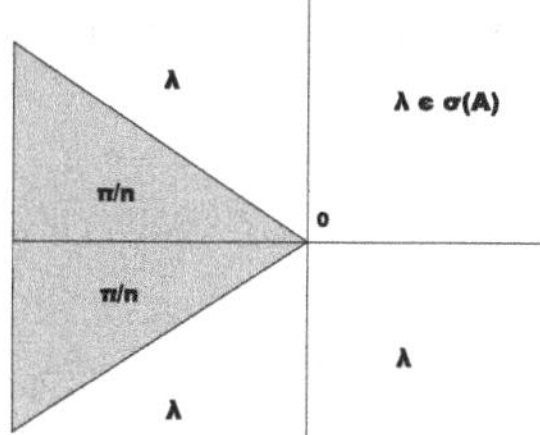

Figure 4.1 The spectrum of a Q-matrix cannot intersect the shaded sector about the negative real axis.

positive because they are the products of their respective eigenvalues, which are either positive or come in complex conjugate pairs. □

Although the real eigenvalues of a P-matrix $A \in M_n(\mathbb{C})$ are positive, other eigenvalues of A may lie in the left half-plane. In fact, as shown in [Her83], P-matrices may have all but one (when n is even), or all but two (when n is odd) of their eigenvalues in the left half-plane.

There is, however, a notable restriction on the location of the eigenvalues of a P-matrix presented in Theorem 4.4.4 and illustrated in the Figure 4.1. It is attributed to Kellogg [Kel72], and it regards the class of matrices whose characteristic polynomials have real coefficients with alternating signs.

Definition 4.4.3 Matrix $A \in M_n(\mathbb{C})$ is called a **Q-matrix** if for each $k \in \langle n \rangle$, the sum of all $k \times k$ principal minors of A is positive. Every P-matrix is indeed a Q-matrix. (This concept of a Q-matrix differs from the one used in the Linear Complementarity Problem literature; see Figure 4.1.)

Theorem 4.4.4 Let $A \in M_n(\mathbb{C})$ be a Q-matrix, and let $\text{Arg}(z) \in (-\pi, \pi]$ denote the principal value of $z \in \mathbb{C}$. Then

$$\sigma(A) \subset \left\{ z \in \mathbb{C} : \ |\,\text{Arg}(z)\,| < \pi - \frac{\pi}{n} \right\}. \tag{4.4.1}$$

Remark 4.4.5 A graph-theoretic refinement of the above result is shown in [JOTvdD93]: If $A \in \mathbf{P}_n^k$, where $k < n$, then

$$\sigma(A) \subset \left\{ z \in \mathbb{C} : \ -\pi + \frac{\pi}{n-1} < \text{Arg}(z) < \pi - \frac{\pi}{n-1} \right\}.$$

Some related refinements were conjectured in [HB83] and studied in [HJ86b] and [Fang89].

Recall that some of the better known subclasses of **P**, like the positive definite matrices and the totally positive matrices, are invariant under the taking of powers, and the eigenvalues of their members are positive numbers. Consequently, it is natural to ask:

Question 4.4.6 Are the eigenvalues of a matrix all of whose powers are P-matrices necessarily positive?

Question 4.4.6 is posed in [HJ86b], where it is associated with spectral properties of Q-matrices. More specifically, the following questions about complex matrices are posed in [HJ86b]:

Question 4.4.7 Is every matrix all of whose powers are Q-matrices co-spectral to a matrix all of whose powers are P-matrices?

Question 4.4.8 Is every Q-matrix similar to a P-matrix?

Question 4.4.9 Is every Q-matrix co-spectral to a P-matrix?

Question 4.4.10 Is every diagonalizable Q-matrix similar to a P-matrix?

Another related question is posed in [HK03, question 6.2]:

Question 4.4.11 Do the eigenvalues of a matrix A such that A and A^2 are P-matrices necessarily have positive real parts?

At the time of writing this book, Questions 4.4.6–4.4.11 remain unanswered. Two results in the direction of resolving these questions are paraphrased below.

Proposition 4.4.12 [TT18] Let $A \in M_n(\mathbb{R})$ such that $tA^2 + (1-t)A \in \mathbf{P}$ for every $t \in [0, 1]$. Then A is positive stable.

Proposition 4.4.13 [Kus16] Let $A \in M_n(\mathbb{R})$ be such that for some permutation matrix P, the leading principal submatrices of PAP^T and the squares of the submatrices are P-matrices. Then A is a positive stable P-matrix.

We conclude this section with a classic result on "diagonal stabilization" of matrices with positive leading principal minors (and thus P-matrices); see Fischer and Fuller [FF58] and Ballantine [Bal70].

Theorem 4.4.14 Let $A \in M_n(\mathbb{R})$ such that $\det A[\langle k\rangle] > 0$ for all $k = 1, 2, \dots, n$. Then there exists a diagonal matrix D with positive diagonal entries such that all the eigenvalues of DA are positive and simple.

4.5 Special Classes of P-Matrices

Below is a list of propositions tantamount to methods that can be used to generate P-matrices with special properties.

Proposition 4.5.1 Every row diagonally dominant matrix $A \in M_n(\mathbb{R})$ with positive diagonal entries is a positive stable P-matrix.

Proof Every principal submatrix $A[\alpha]$ is row diagonally dominant with positive diagonal entries, and thus, by Gershgorin's theorem, $A[\alpha]$ is positive stable. In particular, every real eigenvalue of $A[\alpha]$ is positive. As the complex eigenvalues come in conjugate pairs, it follows the $\det A[\alpha] > 0$ for all nonempty $\alpha \subset \langle n \rangle$. □

Proposition 4.5.2 Let $A \in M_n(\mathbb{R})$ be such that $A + A^T$ is positive definite. Then A is a positive stable P-matrix.

Proof Every principal submatrix $A[\alpha] + A[\alpha]^T$ of $A + A^T$ is also positive definite. Thus for every $x \in \mathbb{C}^{|\alpha|}$,

$$x^* A[\alpha]\, x = x^* \frac{A[\alpha] + A[\alpha]^T}{2} x + x^* \frac{A[\alpha] - A[\alpha]^T}{2} x$$

has positive real part. It follows that every eigenvalue of $A[\alpha]$ has positive real part, and thus every principal submatrix of A is positive stable. As in the proof of Proposition 4.5.1, this implies that A is a positive stable P-matrix. □

Proposition 4.5.3 [Pen01] Let $A \in M_n(\mathbb{R})$ be a B-matrix. Then $A \in \mathbf{P}$.

Proposition 4.5.4 For every nonsingular matrix $A \in M_n(\mathbb{C})$, the matrices A^*A, AA^* are positive definite and thus P-matrices.

Proposition 4.5.5 Given a square matrix $B \geq 0$ and $s > \rho(B)$, $A = sI - \rho(B) \in \mathbf{P}$. In particular, A is an M-matrix and thus positive stable.

Proof Consider $B = [b_{ij}] \geq 0$ and $s > \rho(B)$. By [HJ13, corollary 8.1.20], $s > \rho(B[\alpha])$ for every nonempty $\alpha \subseteq \langle n \rangle$. Thus every principal submatrix $A[\alpha]$ of A is positive stable, and as in the proof of Proposition 4.5.1, it must have positive determinant. □

Proposition 4.5.6 Let $B, C \in M_n(\mathbb{R})$ be row diagonally dominant matrices with positive diagonal entries. Then $BC^{-1} \in \mathbf{P}$.

Proof Let B and C be as prescribed, and notice that by the Levy–Desplanques theorem (see [HJ13, corollary 5.6.17]), $TB + (I - T)C$ is invertible for every $T \in [0, I]$. Thus, by Theorem 4.3.9, $BC^{-1} \in \mathbf{P}$. □

Remark 4.5.7 It is easy to construct examples showing that the above result fails to be true when B or C are not real matrices.

We continue with two results on mimes from [Pan79a]. The original proofs of these results in [Pan79a] rely on (hidden) Leontief matrices and principal pivot transforms. In particular, the proof of Proposition 4.5.9 below is attributed to Mangasarian [Man78], who also used ideas from mathematical programming. We include here shorter proofs that are based on what is considered standard M-matrix and P-matrix theory.

Proposition 4.5.8 Let $A \in M_n(\mathbb{R})$ be a mime. Then $A \in \mathbf{P}$.

Proof Let A be a mime and s_1, s_2, P_1, P_2 and $u \geq 0$ be as in Definition 4.2.3. Then, for every $T \in [0, I]$, the matrix

$$C = T(s_2 I - P_2) + (I - T)(s_1 I - P_1)$$

is a Z-matrix and $Cu > 0$ (i.e., C is a semipositive Z-matrix). This means that C is an M-matrix. In particular, C is invertible for every $T \in [0, I]$. By Theorem 4.3.9, we conclude that $A = (s_1 I - P_1)(s_2 I - P_2)^{-1} \in \mathbf{P}$. □

Proposition 4.5.9 Let $B \geq 0$ with $\rho(B) < 1$. Let $\{a_k\}_{k=1}^m$ be a sequence such that $0 \leq a_{k+1} \leq a_k \leq 1$ for all $k = 1, \ldots, m-1$. Then

$$A = I + \sum_{k=1}^{m} a_k B^k \ \in \mathbf{P}.$$

If m is infinite, under the additional assumption that $\sum_{k=1}^{\infty} a_k$ is convergent, we can still conclude that $A \in \mathbf{P}$. More specifically, A is a nonnegative matrix and a mime.

Proof Consider the matrix $C = A(I - B)$ and notice that C can be written as

$$C = I - G, \ \text{ where } G \equiv B - \sum_{k=1}^{m} a_k \left(B^k - B^{k+1}\right).$$

First we show that G is nonnegative: Indeed, as $0 \leq a_{k+1} \leq a_k \leq 1$, we have

$$\begin{aligned} G &\geq B - \sum_{k=1}^{m} a_k B^k + \sum_{k=1}^{m-1} a_{k+1} B^{k+1} + a_m B^{m+1} \\ &= B - \sum_{k=1}^{m} a_k B^k + \sum_{k=2}^{m} a_k B^k + a_m B^{m+1} \\ &= B - a_1 B + a_m B^{m+1} = (1 - a_1)B + a_m B^{m+1} \geq 0. \end{aligned}$$

Next we show that $\rho(G) < 1$. For this purpose, consider the function

$$\begin{aligned} f(z) &= z - \sum_{k=1}^{m} a_k \left(z^k - z^{k+1}\right) \\ &= z(1 - a_1) + z^2(a_1 - a_2) + \cdots + z^m(a_{m-1} - a_m) + a_m z^{m+1}, \end{aligned}$$

in which expression all the coefficients are by assumption nonnegative. Thus $|f(z)| \leq f(|z|)$. However, for $|z| < 1$, we have

$$f(|z|) = |z| - \sum_{k=1}^{m} a_k \left(|z|^k - |z|^{k+1}\right) \leq |z|.$$

That is, for all $|z| < 1$,

$$|f(z)| \leq f(|z|) \leq |z|.$$

We can now conclude that for every $\lambda \in \sigma(B)$,

$$|f(\lambda)| \leq |\lambda| \leq \rho(B) < 1;$$

that is, $\rho(G) < 1$. We have thus shown that

$$A = (I - G)(I - B)^{-1},$$

where $B, G \geq 0$ and $\rho(B), \rho(G) < 1$. Also, as $B \geq 0$, we may consider its Perron vector $u \geq 0$. By construction, u is also an eigenvector of G corresponding to $\rho(G)$. That is, there exists vector $u \geq 0$ such that

$$Bu = \rho(B)\, u < u \quad \text{and} \quad Gu = \rho(G)\, u < u.$$

Thus A is a mime and so it belongs to $\mathbf{P}$ by Proposition 4.5.8. □

Proposition 4.5.10 Let $B \geq 0$. Then $e^{tB} \in \mathbf{P}$ for all $t \in [0, 1/\rho(B))$.

Proof It follows from Proposition 4.5.9 by taking $m = \infty$ and $a_k = \frac{1}{k!}$. □

H-matrices with positive diagonal entries are mimes and thus P-matrices by the results in [Pan79a]. We include a short direct proof of this fact next.

Proposition 4.5.11 Let $A \in M_n(\mathbb{R})$ be an H-matrix with positive diagonal entries. Then $A \in \mathbf{P}$.

Proof As $\mathcal{M}(A)$ is an M-matrix, by [BP94, theorem 2.3 (M_{35}), chapter 6], there exists a diagonal matrix $D \geq 0$ such that AD is strictly row diagonally dominant. As AD has positive diagonal entries, the result follows from Theorem 4.3.2 (4) and Proposition 4.5.1. □

We now turn our attention to P-matrices generated via sign patterns.

Proposition 4.5.12 $S_n^k \subseteq \mathbf{P}_n^k \subseteq \mathbf{P}$.

Proof Recall that by the definitions in Section 4.2.3, when $A = [a_{ij}] \in S_{n,k}$, the longest cycle in the directed graph of A is $k \geq 1$; the cycles of odd (resp., even) length in the signed directed graph of A are positive (resp., negative).

In particular, all the diagonal entries of A are positive. The terms in the standard expansion of $\det A$ are of the form

$$(-1)^{\text{sign}(\sigma)}\ a_{1\sigma(1)}\ \cdots\ a_{n\sigma(n)}, \tag{4.5.1}$$

where σ is a member of the symmetric permutation group on n elements. However, σ can be uniquely partitioned into its permutation cycles, some of odd and some of even length. In fact, $(-1)^{\text{sign}(\sigma)} = (-1)^{n+q}$, where q is the number of odd cycles. As a consequence, the quantity in (4.5.1) is nonnegative. One such term in $\det A$ consists of the diagonal entries of A, which are positive, and thus the sign of that term is $(-1)^{2n}$. It follows that $\det A > 0$. Notice also that a similar argument can be applied to every principal minor of A. Thus A belongs to $\mathbf{P}_{n,k}$. □

Remark 4.5.13 The argument in the proof above is essentially contained in the proof of [EJ91, theorem 1.9]. Also, notice that the matrices in $S_{n,k}$ are sign nonsingular and sometimes referred to as **qualitative P-matrices**.

Example 4.5.14 The sign pattern

$$\begin{bmatrix} + & + & + \\ - & + & + \\ 0 & - & + \end{bmatrix}$$

belongs to $S_{3,2}$, and thus every matrix with this sign pattern is a P-matrix.

Proposition 4.5.15 Let $A \in M_n(\mathbb{R})$ be sign nonsingular and $B \in M_n(\mathbb{R})$ any matrix with the same sign pattern as A. Then $BA^{-1}, B^{-1}A \in \mathbf{P}$.

Proof Because A and B are sign-nonsingular having the same sign pattern, we have that $C = TB + (I - T)A$ is also sign-nonsingular for every $T \in [0, I]$. Thus, by Theorem 4.3.9, $BA^{-1} \in \mathbf{P}$. The conclusion for $B^{-1}A$ follows from a result in [JT95] dual to the one quoted in Theorem 4.3.9. □

Example 4.5.16 The sign pattern

$$\begin{bmatrix} - & - & 0 & - \\ + & - & - & 0 \\ 0 & + & - & - \\ + & 0 & + & - \end{bmatrix}$$

is sign nonsingular. Thus, by Proposition 4.5.15, for A and B with this sign pattern given by,

$$A = \begin{bmatrix} -1 & -1 & 0 & -1 \\ 1 & -1 & -1 & 0 \\ 0 & 1 & -1 & -2 \\ 3 & 0 & 1 & -2 \end{bmatrix}, \quad B = \begin{bmatrix} -1 & -2 & 0 & -1 \\ 1 & -1 & -3 & 0 \\ 0 & 2 & -2 & -1 \\ 1 & 0 & 1 & -1 \end{bmatrix},$$

we have that

$$BA^{-1} = \frac{1}{23} \begin{bmatrix} 33 & 7 & -6 & 1 \\ -8 & 45 & -14 & -10 \\ -21 & 6 & 31 & -9 \\ -9 & -4 & -13 & 6 \end{bmatrix} \in \mathbf{P}.$$

Remark 4.5.17 Some sufficient conditions for a matrix A to be a P-matrix based on its signed directed graph are provided in [JNT96]. In particular, if $A = [a_{ij}]$ is sign symmetric (i.e., $a_{ij}a_{ji} \geq 0$) and if its undirected graph is a tree (i.e., a connected acyclic graph), then necessary and sufficient conditions that A be a P-matrix are obtained.

4.5.1 Matrices All of Whose Powers Are P-Matrices

As is well known, if A is positive definite, then so are all of its powers. Thus positive definite matrices belong to $\mathbf{P}_{\mathrm{M}}$. By the Cauchy–Binet formula for the determinant of the product of two matrices (see [HJ13]), it follows that powers of totally positive matrices are totally positive and thus belong to $\mathbf{P}_{\mathrm{M}}$. Below is a constructive characterization of totally positive matrices found in [GP96] (see also [Fal01] and [FJ11]), which therefore allows us to generate matrices in $\mathbf{P}_{\mathrm{M}}$.

Theorem 4.5.18 Matrix $A \in M_n(\mathbb{R})$ is totally positive if and only if there exist positive diagonal matrix D and positive numbers $l_i,\ u_j\ (i,j = 1, 2, \ldots, k)$, $k = \binom{n}{2}$, such that $A = FDG$, where

$$F = \left[E_n(l_k)\, E_{n-1}(l_{k-1}) \ \ldots \ E_2(l_{k-n+2})\right] \\ \times \left[E_n(l_{k-n+1}) \ \ldots \ E_3(l_{k-2n+4})\right] \ldots \left[E_n(l_1)\right]$$

and

$$G = \left[E_n^T(u_1)\right] \left[E_{n-1}^T(u_2)\, E_n^T(u_3)\right] \ldots \left[E_2^T(u_{k-n+2}) \ \ldots \ E_{n-1}^T(u_{k-1}) E_n^T(u_k)\right];$$

here $E_k(\beta) = I + \beta E_{k,k-1}$, where $E_{k,k-1}$ denotes the $(k, k-1)$ elementary matrix.

Another subclass of $\mathbf{P}_\mathrm{M}$ is the MMA-matrices, introduced in [FHS87] (recall that these are matrices all of whose positive integer powers are irreducible M-matrices). The results in [ENN98] (see also [Stu98]), combined with the notion of a Soules matrix, allow us to construct all symmetric (inverse) MMA-matrices:

Theorem 4.5.19 Let $A \in M_n(\mathbb{R})$ be an invertible and symmetric matrix. Then A is an MMA-matrix if and only if $A^{-1} = R\Lambda R^T$, where R is a Soules matrix and $\Lambda = \mathrm{diag}(\lambda_1, \ldots, \lambda_n)$ with $\lambda_1 > \lambda_2 \geq \cdots \geq \lambda_n > 0$.

It has also been shown in [FHS87] that for every MMA-matrix $B \in M_n(\mathbb{R})$ there exists a positive diagonal matrix D such that $A = D^{-1}BD$ is a symmetric MMA-matrix. In [HS86], D is found to be

$$D = \mathrm{diag}\left(x_1^{1/2} y_1^{-1/2}, x_2^{1/2} y_2^{-1/2}, \ldots, x_n^{1/2} y_n^{-1/2}\right),$$

where x, y are the unit strictly positive right and left Perron eigenvectors of B, respectively. Thus, we have a way of constructing arbitrary MMA-matrices as follows: Determine an orthogonal Soules matrix R starting with an arbitrary unit positive vector w_1. Choose a matrix Λ as in Theorem 4.5.19 and form $R\Lambda R^T$. Then $A = R\Lambda^{-1}R^T$ is a symmetric MMA-matrix with unit left and right Perron eigenvectors equal to w_1. Choose now positive diagonal matrix $\hat{D}$ and let $B = \hat{D}R\Lambda^{-1}R^T\hat{D}^{-1}$. Then B is an MMA-matrix having $\hat{D}w_1$ and $\hat{D}^{-1}w_1$ are right and left Perron vectors, respectively.

Remark 4.5.20 Given a matrix A in $\mathbf{P}_\mathrm{M}$, so are A^T, $D^{-1}AD$ (where D is a positive diagonal matrix), QAQ^T (where Q is a permutation matrix), and SAS (where S is a signature matrix).

4.5.2 More on Mimes

It has been shown in [Pan79a] that the notion of a mime A as in Definition 4.2.3 is tantamount to A being "hidden Z" and a P-matrix at the same time. More relevant to our context is the following result, reproven here using the language and properties of M-matrices.

Proposition 4.5.21 Let $A \in M_n(\mathbb{R})$. Then A is a mime if and only if
(1) $AX = Y$ for some Z-matrices X and Y, and
(2) A and X are semipositive.

Proof Clearly, if A is a mime as in Definition 4.2.3, then (1) holds with the roles of X and Y being played by $(s_2I - P_2)$ and $(s_1I - P_1)$, respectively. That

(2) holds follows from the fact that $z = Xu > 0$ (i.e., X is semipositive) and $Yu > 0$, where $u \geq 0$ is a common semipositivity vector of A. We then have that $Az = YX^{-1}Xu = Yu > 0$; that is, A is also semipositive.

For the converse, suppose (1) and (2) hold. Then X is an invertible M-matrix. As A is assumed semipositive, $Ax > 0$ for some $x \geq 0$. Let $u = X^{-1}x$. Then, $Yu = Ax > 0$; that is, Y is also semipositive and so an M-matrix as well. In fact, u is a common semipositivity vector of A and thus A is a mime. □

Remarks 4.5.22

(i) Based on the above result, we can assert that given a mime A and a positive diagonal matrix D, $A + D$ and DA are also mimes.

(ii) Principal pivot transforms, permutation similarities, Schur complementation, and extraction of principal submatrices leave the class of mimes invariant; see [Pan79a, Tsa00].

Recall that the mimes form a subclass of the P-matrices that includes the M-matrices, the H-matrices with positive diagonal entries, as well as their inverses. The following is another large class of mimes mentioned in [Pan79b].

Theorem 4.5.23 Every triangular P-matrix is a mime.

Proof We prove the claim by induction on the order k of A. If $k = 1$ the result is obviously true. Assume the claim is true for $k = n - 1$; we will prove it for $k = n$. For this purpose, consider the triangular P-matrix

$$A = \begin{bmatrix} A_{11} & a \\ 0 & a_{22} \end{bmatrix},$$

where A_{11} is an $(n-1)$-by-$(n-1)$ P-matrix, $a \in \mathbb{R}^{n-1}$ and $a_{22} > 0$. By the inductive hypothesis and Proposition 4.5.21, there exist Z-matrices X_{11}, Y_{11} and nonnegative vector $u_1 \in \mathbb{R}^{n-1}$ such that $A_{11}X_{11} = Y_{11}$, $Y_{11}u_1 > 0$, and $X_{11}u_1 > 0$. Consider then the Z-matrix

$$X = \begin{bmatrix} X_{11} & -X_{11}u_1 \\ 0 & x_{22} \end{bmatrix},$$

where $x_{22} > 0$ is to be chosen. Then let

$$Y = AX = \begin{bmatrix} A_{11}X_{11} & -Y_{11}u_1 + x_{22}a \\ 0 & a_{22}x_{22} \end{bmatrix}.$$

Notice that $x_{22} > 0$ can be chosen so that Y is a Z-matrix. Let also $u^T = \begin{bmatrix} u_1^T & u_2 \end{bmatrix}$. Choosing $u_2 > 0$ small enough, we have that $Xu > 0$ and $Yu > 0$. Thus A is a mime by Proposition 4.5.21. □

Theorem 4.5.24 Let A be a mime. Then A can be factored into $A = BC^{-1}$, where B, C are row diagonally dominant matrices with positive diagonal entries.

Proof Suppose that $A = (s_1 I - P_1)(s_2 I - P_2)^{-1}$ is a mime with semipositivity vector u that by continuity can be chosen to be positive ($u > 0$). Let $D = \text{diag}(u_1, \ldots, u_n)$. Define $B = (s_1 I - P_1)D$ and $C = (s_2 I - P_2)D$ so that $Be > 0$ and $Ce > 0$, where e is the all ones vector. Notice that as B and C are Z-matrices with positive diagonal entries, they are indeed row diagonally dominant, and $A = BC^{-1}$. □

Remark 4.5.25 Not all P-matrices are mimes as shown by the construction of a counterexample in Pang [Pan79b].

4.6 The P-Problem: Detection of P-Matrices

Of particular concern in applications is the ability to decide, as efficiently as possible, whether an n-by-n matrix is in **P** or not. This is referred to as the **P-problem**. This problem has received attention due to its inherent computational complexity, as well as due to the connection of P-matrices to the linear complementarity problem (see Section 4.9.1) and to self-validating methods for its solution (see [CSY01, JR99, RR96, Rum01]).

The P-problem is indeed NP-hard. Specifically, the P-problem has been shown by Coxson [Cox94, Cox99] to be co-NP-complete. That is, the time complexity of discovering that a given matrix has a negative principal minor is NP-complete.

Checking the sign of each principal minor is indeed a task of exponential complexity. A strategy has been developed in [Rum03] for detecting P-matrices, which is not *a priori* exponential; however, it can be exponential in the worst case.

The most efficient to date comprehensive method for the P-problem is a recursive $O(2^n)$ algorithm developed in [TL00], which is simple to implement and lends itself to computation in parallel; it is presented in detail in Section 4.6.2.

For certain matrix classes discussed in Section 4.6.1, the P-problem is computationally straightforward, in some cases demanding only low-degree polynomial time complexity.

4.6.1 Detecting Special P-Matrices

To detect a positive definite matrix A one needs to check if $A = A^*$ and whether the eigenvalues of A are positive or not. To detect an M-matrix A, one needs to check if A is a Z-matrix and whether A is positive stable or not. An alternative $O(n^3)$ procedure for testing if a Z-matrix is an M-matrix is described in [Val91]. To detect an inverse M-matrix A, one can apply a test for M-matrices to A^{-1}. Similarly, one needs to apply a test for M-matrices to $\mathcal{M}(A)$ to determine whether a matrix A with positive diagonal entries is an H-matrix or not.

Recall that an H-matrix is also characterized by the existence of a positive diagonal matrix D such that AD is row diagonally dominant. This is why an H-matrix is also known in the literature as a **generalized diagonally dominant matrix**. Based on this characterization of H-matrices, an iterative method to determine whether A is an H-matrix or not is developed in [LLHNT98]; this iterative method also determines a diagonal matrix D with the aforementioned scaling property. More methods to recognize H-matrices have since been developed, see, e.g., [AH08].

To determine a totally positive matrix, Fekete's criterion [FP1912] may be used: Let $S = \{i_1, i_2, \dots, i_k\}$ with $i_j < i_{j+1}, \ (j = 1, 2, \dots, k-1)$, and define the **dispersion** of S to be

$$d(S) = \begin{cases} i_k - i_1 - k + 1 & \text{if } k > 1 \\ 0 & \text{if } k = 1. \end{cases}$$

Then $A \in M_n(\mathbb{R})$ is totally positive if and only if $\det A[\alpha|\beta] > 0$ for all $\alpha, \beta \subseteq \langle n \rangle$ with $|\alpha| = |\beta|$ and $d(\alpha) = d(\beta) = 0$. That is, one only needs to check for positivity minors whose row and column index sets are contiguous. Fekete's criterion is improved in [GP96] as follows:

Theorem 4.6.1 Matrix $A \in M_n(\mathbb{R})$ is totally positive if and only if for every $k \in \langle n \rangle$

(a) $\det A[\alpha|\langle k \rangle] > 0$ for all $\alpha \subseteq \langle n \rangle$ with $|\alpha| = k$ and $d(\alpha) = 0$,
(b) $\det A[\langle k \rangle|\beta] > 0$ for all $\beta \subseteq \langle n \rangle$ with $|\beta| = k$ and $d(\beta) = 0$.

A complete discussion for the recognition of total positivity can be found in [FJ11, chapter 3].

The task of detecting mimes is less straightforward and can be based on Proposition 4.5.21. As argued in [Pan79b] and by noting that, without loss of any generality, the semipositivity vector of X in Proposition 4.5.21 can be taken to be the all ones vector e (otherwise, our considerations apply with X, Y

replaced by XD, YD for a suitable positive diagonal matrix D), the following two-step test for mimes is applicable [Pan79b]:

Step 1. Determine whether A is semipositive or not by solving the linear program

```
minimize e^T x
subject to x ≥ 0 and Ax ≥ e
```

If this program is infeasible, then A is **not** semipositive and thus not a mime; stop.

Step 2. Check the consistency of the linear inequality system

$$x_{ij} \le 0 \quad (i,j = 1,2,\ldots,n,\ i \ne j),$$
$$\textstyle\sum_{k=1}^{n} a_{ik}x_{kj} \le 0 \quad (i,j = 1,2,\ldots,n,\ i \ne j),$$
$$\textstyle\sum_{j=1}^{n} x_{ij} > 0 \quad (i = 1,2,\ldots,n).$$

If this system is inconsistent, A is not a mime; stop. Otherwise, A is a mime for it satisfies the conditions of Proposition 4.5.21.

Remark 4.6.2 The task of detecting matrices in $\mathbf{P}_{\mathrm{M}}$ is an open problem, related to the problem of characterizing matrices in $\mathbf{P}_{\mathrm{M}}$ and their spectra; see Section 4.4 of this chapter.

4.6.2 Detection of General P-Matrices

The exhaustive check of all $2^n - 1$ principal minors of $A \in M_n(\mathbb{R})$ using Gaussian elimination is an $O(n^3 2^n)$ task; see [TL00]. Next we will describe an alternative test for complex P-matrices, which was first presented in [TL00] and was shown to be $O(2^n)$ when applied to real P-matrices. The following theorem (cf. Theorem 4.3.2 (7) and (8)) is the theoretical basis for the subsequent algorithm. Although the result is stated and proved in [TL00] for real P-matrices, its proof is valid for *complex* P-matrices and included here for completeness.

Theorem 4.6.3 Let $A \in M_n(\mathbb{C})$, $\alpha \subseteq \langle n \rangle$ with $|\alpha| = 1$. Then $A \in \mathbf{P}$ if and only if $A[\alpha] > 0$, $A[\overline{\alpha}] \in \mathbf{P}$, and $A/A[\alpha] \in \mathbf{P}$.

Proof Without loss of generality, assume that $\alpha = \{1\}$. Otherwise we can apply our considerations to a permutation similarity of A. If $A = [a_{ij}] \in \mathbf{P}$, then $A[\alpha] \in \mathbf{P}$ and $A[\overline{\alpha}] \in \mathbf{P}$. Also $A/A[\alpha] \in \mathbf{P}$ by Theorem 4.3.2 (7). For the converse, assume that $A[\alpha] = [a_{11}]$, $A[\overline{\alpha}]$, and $A/A[\alpha]$ are P-matrices. Using

$a_{11} > 0$ as the pivot, we can row reduce A to obtain a matrix B with all of its off-diagonal entries in the first column equal to zero. As is well known, $B[\overline{\alpha}] = A/A[\alpha]$. That is, B is a block triangular matrix whose diagonal blocks are P-matrices. It follows readily that $B \in \mathbf{P}$. The determinant of any principal submatrix of A that includes entries from the first row of A coincides with the determinant of the corresponding submatrix of B and is thus positive. The determinant of any principal submatrix of A with no entries from the first row coincides with a principal minor of $A[\overline{\alpha}]$ and is also positive. Hence $A \in \mathbf{P}$. □

The following is an implementation of a recursive algorithm for the P-problem suggested by Theorem 4.6.3. For its complexity analysis, see [GT06a, TL00].

ALGORITHM P(A)

1. Input $A = [a_{ij}] \in M_n(\mathbb{C})$.
2. If $a_{11} \not> 0$ output **"Not a P-matrix"** stop.
3. Evaluate $A/A[\alpha]$, where $\alpha = \{1\}$.
4. Call P($A(\alpha)$) and P($A/A[\alpha]$).
5. Output **"This is a P-matrix."**

Next is a Matlab function implementing Algorithm P(A) for general complex matrices (see the remark below for online availability).

```
function [r] = ptest(A)
% Return r=1 if 'A' is a P-matrix (r=0 otherwise).
  n = length(A);
if ~(A(1,1)>0), r = 0;
elseif n==1, r = 1;
else
   b = A(2:n,2:n);
   d = A(2:n,1)/A(1,1);
   c = b - d*A(1,2:n);
   r = ptest(b) & ptest(c);
end
```

Remark 4.6.4 The algorithm P(A) uses Schur complementation and submatrix extraction in a recursive manner to compute (up to) $2n$ quantities. In the course of P(A), if any of these quantities is not positive, the algorithm terminates, declaring that A is not a P-matrix; otherwise it is a P-matrix. No further use of these $2n$ quantities is made in P(A), even when they all have to be computed; they are in fact overwritten. In [GT06a] an algorithm is developed (MAT2PM) that uses the underlying technique and principles of P(A) in order to compute all the principal minors of an arbitrary complex matrix. In [GT06b], a method is presented (PM2MAT) that constructs recursively a matrix from a set of prescribed principal minors, when this is possible. Matlab implementations of the algorithms in [GT06a, GT06b], as well as ptest, are maintained in `www.math.wsu.edu/math/faculty/tsat/matlab.html`.

4.7 The Recursive Construction of All P-Matrices

The results in this section are based on material introduced in [TZ19]. Of particular interest in the study of P-matrices are two related problems:

(i) Recognize whether or not a given matrix is a P-matrix (P-problem; see Section 4.6).
(ii) Provide a constructive characterization of P-matrices, namely, a method that can generate every P-matrix.

Both problems are central to computational challenges arising in the Linear Complementarity Problem (LCP) (see Section 4.9.1) and exemplified by the following facts:

- LCP has a unique solution if and only if the coefficient matrix is a P-matrix [CPS92].
- The problem of deciding if a given matrix is a P-matrix is co-NP-complete [Cox94].
- The complexity of solving the LCP when the coefficient matrix is a P-matrix is presently unknown. If the problem of solving the LCP with a P-matrix were NP-hard, then the complexity classes NP and co-NP would coincide [MP91].

One interesting method to construct real P-matrices is to form products $A = BC^{-1}$, where $B, C \in M_n(\mathbb{R})$ are row diagonally dominant matrices with positive diagonal entries (see Proposition 4.5.6). This was first observed in [JT95] and is also the subject of study in [Mor03] and [MN07], where such matrices are referred to as "hidden prdd." At an Oberwolfach meeting [Ober], C. R. Johnson stated the above result and raised the question whether or not all real P-matrices can be factored this way, namely, they are hidden prdd. A counterexample was provided in [Mor03], along with a polynomial algorithm to detect hidden prdd matrices. Further study of related classes was pursued in [MN07].

We shall see that the solution to problem (ii) above is intrinsically related to problem (i). Using a recursion based on rank-one perturbations of P-matrices, we shall be able to reverse the steps of the recursive algorithm P(A) in Section 4.6.2 that detects P-matrices in order to construct every P-matrix.

4.7.1 Rank-One Perturbations in Constructing P-Matrices

The main idea in Theorem 4.6.3 allows one to construct recursively any and all members of the class of P-matrices of any given size. This is presented in the next two theorems and algorithm.

Proposition 4.7.1 Let $\hat{A} \in M_n(\mathbb{C})$ be a P-matrix, $a \in \mathbb{C}$ and $x, y \in \mathbb{C}^n$. Then the following are equivalent.

(i) $A = \begin{bmatrix} a & x^T \\ -y & \hat{A} \end{bmatrix}$ is a P-matrix.

(ii) $a > 0$ and $\hat{A} + \frac{1}{a} yx^T$ is a P-matrix.

Proof The equivalence follows from Theorem 4.6.3 and the fact that $\hat{A} + \frac{1}{a} yx^T$ is the Schur complement of $a = A[\{1\}]$ in A. □

Proposition 4.7.1 suggests the following recursive process to construct an $n \times n$ complex P-matrix, $n \geq 2$.

ALGORITHM `P-CON`

Choose $A_1 > 0$
For $k = 1 : n - 1$, given the $k \times k$ P-matrix A_k

1. Choose $(x^{(k)}, y^{(k)}) \in \mathbb{C}^k \times \mathbb{C}^k$ and $a_k > 0$ such that $A_k + \frac{1}{a_k} y^{(k)} \left(x^{(k)}\right)^T$ is a P-matrix.
2. Form the $(k+1) \times (k+1)$ matrix $A_{k+1} = \begin{bmatrix} a_k & \left(x^{(k)}\right)^T \\ -y^{(k)} & A_k \end{bmatrix}$.

Output $A = A_n$ **is a P-matrix**.

The recursive nature of every P-matrix is formally shown in the following theorem.

Theorem 4.7.2 Every matrix constructed via `P-CON` is a P-matrix. Conversely, every P-matrix $A \in M_n(\mathbb{C})$ can be constructed via `P-CON`.

Proof By Proposition 4.7.1, each of the matrices A_{k+1} ($k = 1, 2, \ldots, n-1$) in `P-CON`, including A_1, is a P-matrix. We use induction to prove the converse.

The base case is trivial. Assume every P-matrix in $M_{n-1}(\mathbb{C})$ can be constructed via P-CON. Let $A \in M_n(\mathbb{C})$ be any P-matrix partitioned as

$$A = \begin{bmatrix} a_{11} & A_{12} \\ A_{21} & A_{22} \end{bmatrix},$$

where $A_{22} \in M_{n-1}(\mathbb{C})$. By inductive hypothesis, A_{22} is a P-matrix constructible via P-CON. Because A is a P-matrix, by Corollary 4.7.1, $A/a_{11} = A_{22} - \frac{1}{a_{11}}A_{21}A_{12}$ is a P-matrix, and $A_n = A$ is constructible via P-CON with

$$a_{n-1} = a_{11} > 0,\ x^{(n-1)} = A_{12}^T \text{ and } y^{(n-1)} = -A_{21}. \qquad \square$$

4.7.2 Constructing P-Matrices

Our further development and application of algorithm P-CON in this section will be guided by the following considerations.

(1) The choice of $\left(x^{(k)}, y^{(k)}\right) \in \mathbb{C}^k \times \mathbb{C}^k$ and $a_k > 0$ in P-CON must be made such that $A_k + \frac{1}{a_k}y^{(k)}\left(x^{(k)}\right)^T$ has positive principal minors. Given that the primary interest in applications concerns real P-matrices, we will offer an implementation of P-CON that constructs real P-matrices. Therefore, we execute Step 1 of P-CON by choosing a pair of real vectors $\left(x^{(k)}, y^{(k)}\right)$ randomly, and subsequently choose $a_k > 0$ sufficiently large to ensure $A_k + \frac{1}{a_k}y^{(k)}\left(x^{(k)}\right)^T$ is a P-matrix.

(2) Theorem 4.7.2 and P-CON deal with complex P-matrices. To proceed with construction of non-real P-matrices, it is implicit in the condition that $A_k + \frac{1}{a_k}y^{(k)}\left(x^{(k)}\right)^T$ be a P-matrix that, although the j-th entries of $x^{(k)}$, $y^{(k)}$ can be non-real, their product must be real. We will illustrate such a construction in Example 4.7.7.

Pursuant to consideration (1) above, the lemma and theorem that follow facilitate the choice of a_k in the implementation of P-CON.

Lemma 4.7.3 Let $A \in M_n(\mathbb{C})$ be invertible and $y, x \in \mathbb{C}^n$. Then $\det\left(A + yx^T\right) = \left(x^T A^{-1} y + 1\right) \det(A)$.

Proof Because

$$\begin{bmatrix} I & 0 \\ x^T & 1 \end{bmatrix} \begin{bmatrix} I + yx^T & y \\ 0 & 1 \end{bmatrix} \begin{bmatrix} I & 0 \\ x^T & 1 \end{bmatrix}^{-1} = \begin{bmatrix} I & 0 \\ x^T & 1 \end{bmatrix} \begin{bmatrix} I + yx^T & y \\ 0 & 1 \end{bmatrix} \begin{bmatrix} I & 0 \\ -x^T & 1 \end{bmatrix}$$

$$= \begin{bmatrix} I & y \\ 0 & x^T y + 1 \end{bmatrix},$$

we have

$$\det \begin{bmatrix} I + yx^T & y \\ 0 & 1 \end{bmatrix} = \det \begin{bmatrix} I & y \\ 0 & x^T y + 1 \end{bmatrix},$$

i.e., $\det\left(I + yx^T\right) = x^T y + 1$. Thus,

$$\det\left(A + yx^T\right) = \det(A)\ \det\left(I + A^{-1}yx^T\right) = \left(x^T A^{-1} y + 1\right)\det(A). \qquad \square$$

Theorem 4.7.4 Let $A \in M_n(\mathbb{C})$ be a P-matrix, $y, x \in \mathbb{C}^n$ and $a > 0$. Then $A + \frac{1}{a}yx^T$ is a P-matrix if and only if for every $\alpha \subseteq \langle n \rangle$, we have $(x[\alpha])^T(A[\alpha])^{-1}y[\alpha] \in \mathbb{R}$ and

$$a > \max_{\alpha \subseteq \langle n \rangle}\left\{ -(x[\alpha])^T(A[\alpha])^{-1}y[\alpha]\right\}.$$

Proof For every $\alpha \subseteq \langle n \rangle$, $\det(A[\alpha]) > 0$ because A is a P-matrix. Notice that the principal submatrices of $A + \frac{1}{a}yx^T$ are of the form $A[\alpha] + \frac{1}{a}y[\alpha](x[\alpha])^T$.

First we prove sufficiency: Suppose that for every $\alpha \subseteq \langle n \rangle$, $(x[\alpha])^T(A[\alpha])^{-1}$ $y[\alpha] \in \mathbb{R}$ and

$$a > \max_{\alpha \subseteq \langle n \rangle}\left\{ -(x[\alpha])^T(A[\alpha])^{-1}y[\alpha]\right\}.$$

By Lemma 4.7.3, for any $\alpha \subseteq \langle n \rangle$, we have

$$\det\left(A[\alpha] + \frac{1}{a}y[\alpha](x[\alpha])^T\right) = \left(\frac{1}{a}(x[\alpha])^T(A[\alpha])^{-1}y[\alpha] + 1\right)\det(A[\alpha]) > 0$$

Therefore, $A + \frac{1}{a}yx^T$ is a P-matrix.

Next we prove necessity. If $A + \frac{1}{a}yx^T$ is a P-matrix, then for every $\alpha \subseteq \langle n \rangle$,

$$\det\left(A[\alpha] + \frac{1}{a}y[\alpha](x[\alpha])^T\right) = \left(\frac{1}{a}(x[\alpha])^T(A[\alpha])^{-1}y[\alpha] + 1\right)\det(A[\alpha]) > 0,$$

which necessitates that $(x[\alpha])^T(A[\alpha])^{-1}y[\alpha] \in \mathbb{R}$ and $\frac{1}{a}(x[\alpha])^T(A[\alpha])^{-1}y[\alpha] + 1 > 0$, i.e.,

$$a > \max_{\alpha \subseteq \langle n \rangle}\left\{ -(x[\alpha])^T(A[\alpha])^{-1}y[\alpha]\right\}. \qquad \square$$

Next, we incorporate the bound in Theorem 4.7.4 in Matlab code (PCON) that constructs real P-matrices.

PCON

```
function [A]=pcon(N)
% Input N is the size of the desirable real P-matrix
                                          to be generated
 A=rand(1); % or A=abs(normrnd(0,1)); random 1x1 P-matrix
for j=1:N-1
  x=-1+2.*rand(j,1); % random entries in [-1,1];
                              or use x=normrnd(0,1,[j 1]);
  y=-1+2.*rand(j,1); %        or use y=normrnd(0,1,[j 1]);
  [m,n]=size(A);
  v=1:n;
  a=0.01; % or a=abs(normrnd(0,1));
  for k=1:n
    C = nchoosek(v,k);
    [p,q]=size(C);
    for i=1:p
      B=A(C(i,:), C(i,:));
      b=-(x(C(i,:))).'*inv(B)*y(C(i,:));
      if b>a
        a=b;
      end
    end
  end
  a=1.01*a; % or a=(1+abs(normrnd(0,1)))*a;
  A=[a   x.'; -y  A];
end
```

Remark 4.7.5 The following remarks clarify the functionality of the implementation of P-CON.

- The complexity of P-CON is exponential because of the computation of the lower bound for $a = a_k$ in Theorem 4.7.4, which requires a maximum be computed among all n-choose-k submatrices.
- In the code provided above, random choices are uniformly distributed; normal distribution commands are commended out.
- The lower bound for the parameter a_k provided by Theorem 4.7.4 is strict, hence our choice of a_k is larger than (but kept close to) the lower bound to minimize the chance of diagonal dominance.
- Given that P-matrices are preserved under positive scaling of the rows and columns, there is no loss of generality in restricting the (uniformly distributed) random choice of the entries of $x^{(k)}$ and $y^{(k)}$ to be in $[-1, 1]$.

- We have experimented with normal and uniform distributions for the random choice of the parameters and vector entries. We have, however, observed no discernible difference in the nature of the generated P-matrices.
- As desirable, in the experiments we have run, the matrices generated display no symmetry, no sign pattern, and no diagonal dominance because none of these traits are imposed by `P-CON`.

We conclude by illustrating the functionality of `P-CON` with some generated examples of real P-matrices. We also include an example of a non-real P-matrix generated via `P-CON`.

Example 4.7.6 The first two examples are P-matrices generated by execution of `P-CON` with random variables that are uniformly distributed.

$$\begin{bmatrix} 0.8944 & -0.1366 & 0.9951 & 0.6232 \\ 0.0287 & 0.0101 & -0.7969 & -0.2183 \\ -0.7889 & 0.8908 & 0.0101 & 0.8127 \\ 0.7249 & -0.0026 & -0.2578 & 0.5102 \end{bmatrix},$$

$$\begin{bmatrix} 5.5491 & 0.0613 & 0.6648 & 0.1950 & -0.3294 \\ 0.4015 & 0.4512 & -0.6777 & 0.5162 & 0.7422 \\ 0.0948 & 0.2984 & 40.4328 & -0.1956 & 0.2413 \\ 0.1547 & -0.3711 & 0.6913 & 0.2783 & -0.1952 \\ 0.2808 & 0.4117 & 0.2373 & -0.9657 & 0.6841 \end{bmatrix}.$$

In the next two examples random variables were normally distributed.

$$\begin{bmatrix} 2.7122 & 1.1093 & -0.8637 & 0.0774 \\ 1.2141 & 1.3140 & 0.3129 & -0.8649 \\ 1.1135 & 0.0301 & 0.3185 & -1.7115 \\ 0.0068 & 0.1649 & 0.1022 & 1.3703 \end{bmatrix},$$

$$\begin{bmatrix} 5.7061 & 2.5260 & 1.6555 & 0.3075 & -1.2571 \\ 0.8655 & 1.7977 & -0.1332 & -0.7145 & 1.3514 \\ 0.1765 & 0.2248 & 1.6112 & 0.9642 & 0.5201 \\ -0.7914 & 0.5890 & 0.0200 & 1.6021 & -0.9792 \\ 1.3320 & 0.2938 & 0.0348 & 1.1564 & 0.8314 \end{bmatrix}.$$

Example 4.7.7 In this example, we explicitly illustrate the construction of a 4×4 non-real P-matrix using `P-CON`. The choices are made deliberately to satisfy the necessary conditions and are not randomly generated.

Choose $A_1 = 2$, $x^{(1)} = 2 - i$, $y^{(1)} = 4 + 2i$ and $a_1 = 1$ so that

$$A_1 + \frac{1}{a_1} y^{(1)} \left(x^{(1)}\right)^T = 12$$

is a 1×1 P-matrix. Thus $A_2 = \begin{bmatrix} 1 & \frac{2-i}{2} \\ -4-2i & 2 \end{bmatrix}$ is a P-matrix.

Choose $x^{(2)} = \begin{bmatrix} -1-2i \\ 3-i \end{bmatrix}$, $y^{(2)} = \begin{bmatrix} \frac{1}{2}-i \\ 3+i \end{bmatrix}$ and $a_2 = 3$ so that

$$A_2 + \frac{1}{a_2} y^{(2)} \left(x^{(2)}\right)^T = \begin{bmatrix} 0.1667 & 2.1667 - 2.1667i \\ -4.3333 - 4.3333i & 5.3333 \end{bmatrix}$$

is a P-matrix. Thus $A_3 = \begin{bmatrix} 3 & -1-2i & 3-i \\ -\frac{1}{2}+i & 1 & 2-i \\ -3-i & -4-2i & 2 \end{bmatrix}$ is a P-matrix.

Choose $x^{(3)} = \begin{bmatrix} -i \\ 1+i \\ \frac{1}{3} \end{bmatrix}$, $y^{(3)} = \begin{bmatrix} 2i \\ 1-i \\ -\frac{2}{3} \end{bmatrix}$ and $a_3 = 3.5$ so that

$$A_3 + \frac{1}{a_3} y^{(3)} \left(x^{(3)}\right)^T$$
$$= \begin{bmatrix} 3.5714 & -1.5714 - 1.4286i & 3.0000 - 0.8095i \\ -0.7857 + 0.7143i & 1.5714 & 2.0952 - 1.0952i \\ -3.0000 - 0.8095i & -4.1905 - 2.1905i & 1.9365 \end{bmatrix}$$

is a P-matrix. Thus $A_4 = \begin{bmatrix} 3.5 & -i & 1+i & \frac{1}{3} \\ -2i & 3 & -1-2i & 3-i \\ -1+i & -\frac{1}{2}+i & 1 & 2-i \\ \frac{2}{3} & -3-i & -4-2i & 2 \end{bmatrix}$ is a P-matrix.

4.8 On Matrices with Nonnegative Principal Minors

Many of the properties and characterizations of P-matrices extend to $\mathbf{P}_0$**-matrices**, namely, the class $\mathbf{P}_0$ of matrices whose principal minors are nonnegative. Some of these extensions are straightforward consequences of the definitions and some are based on the following lemma, which implies that $\mathbf{P}_0$ is the topological closure of $\mathbf{P}$ in $M_n(\mathbb{C})$.

Lemma 4.8.1 Let $A \in M_n(\mathbb{C})$. Then $A \in \mathbf{P}_0$ if and only if $A + \epsilon I \in \mathbf{P}$ for all $\epsilon > 0$.

Proof Suppose $A = [a_{ij}] \in \mathbf{P}_0$ so that $\frac{\partial \det A}{\partial a_{ii}} = \det A[\overline{\{i\}}] \geq 0$; that is, $\det A$ is a non-decreasing function of each of the diagonal entries. Thus, as each of the diagonal entries of A is increased in succession by $\epsilon > 0$, the determinant of $A + \epsilon I$ must remain nonnegative. In addition, the coefficients of the characteristic polynomial of A, $\det(tI - A)$, alternate from nonnegative to non-positive and thus does not have any negative roots. That is, A has no negative eigenvalues and so $A + \epsilon I$ is nonsingular, implying that $\det(A + \epsilon I) > 0$. A similar argument applies to every principal submatrix of A, proving that $A + \epsilon I \in \mathbf{P}$ for every $\epsilon > 0$.

Conversely, by continuity of the determinant as a function of the entries, if $A+\epsilon I$ is a P-matrix for every $\epsilon > 0$, as ϵ approaches zero, the principal minors of A must be nonnegative. □

The following facts about $\mathbf{P}_0$-matrices can be obtained from the definitions, as well as the corresponding facts about **P**-matrices and the above lemma.

Theorem 4.8.2 Let $A \in M_n(\mathbb{C})$ and its principal submatrices have real characteristic polynomials. Then A is a P_0-matrix if and only if every real eigenvalue of every principal submatrix of A is nonnegative.

Theorem 4.8.3 Let $A \in M_n(\mathbb{C})$ be a P_0-matrix and let $\mathrm{Arg}z \in (-\pi, \pi]$ denote the principal value of $z \in \mathbb{C}$. Then

$$\sigma(A) \subset \left\{z \in \mathbb{C}: \ |\mathrm{Arg}(z)| \leq \pi - \frac{\pi}{n}\right\}. \tag{4.8.1}$$

Theorem 4.8.4 Let $A \in M_n(\mathbb{C})$ be a P_0-matrix. Then

(1) $A/A[\alpha] \in \mathbf{P}_0$ for all $\alpha \subseteq \langle n \rangle$ such that $A[\alpha]$ is invertible.
(2) $\mathrm{ppt}(A,\alpha) \in \mathbf{P}_0$ for all $\alpha \subseteq \langle n \rangle$ such that $A[\alpha]$ is invertible. In particular, if A is invertible, then $A^{-1} \in \mathbf{P}_0$.

Theorem 4.8.5 $A \in M_n(\mathbb{R})$ is a P_0-matrix if and only if for each $x \in \mathbb{R}^n$, there exists $j \in \langle n \rangle$ such that $x_j(Ax)_j \geq 0$.

4.9 Some Applications of P-Matrices

We provide a brief description of some notable manifestations of P-matrices.

4.9.1 Linear Complementarity Problem

Let $M \in M_n(\mathbb{R})$ and $q \in \mathbb{R}^n$ be given. The **Linear Complementarity Problem**, denoted by LCP (q, M), is to find $z \in \mathbb{R}^n$ such that

$$\begin{aligned} z &\geq 0 \\ q + Mz &\geq 0 \\ z^T(q+Mz) &= 0. \end{aligned}$$

An equivalent formulation of the LCP (q, M) is as the quadratic program

$$\begin{aligned} &\text{minimize} \quad z^T(q+Mz) \\ &\text{subject to} \quad q+Mz \geq 0 \ \text{ and } \ z \geq 0. \end{aligned}$$

Many of the matrix positivity classes discussed herein appear in the Linear Complementarity Problem literature. For a guide to these classes and their roles, see [Cott10]. Most relevant to this chapter is the following theorem, whose proof can be found in [CPS92] or [BP94].

Theorem 4.9.1 The LCP (q, M) has a unique solution for each vector $q \in \mathbb{R}^n$ if and only if $M \in \mathbf{P}$.

Remarks 4.9.2 (i) Related to Theorem 4.9.1, it can be shown that LCP (q, M) has a unique solution for certain $4n + 1$ vectors $q \in \mathbb{R}^n$, then it has unique solution for all q. Hence P-matrices can be characterized by the uniqueness of the solution to a finite number of Linear Complementarity Problems; see [Mu71].

(ii) The solution to the LCP (q, M) can be found by pivoting methods (e.g., Lemke's algorithm) in exponential time, or by iterative methods (e.g., interior point methods), which are better suited for large-scale problems. A comprehensive presentation of numerical methods for the LCP (q, M) , as well as their adaptations to special matrix classes, can be found in [CPS92].

(iii) Several computational issues arise in the solution of the LCP (q, M) [Meg88], which have implications in the development of numerical methods and general complexity theory. For example:

- **[P-problem]** Detect a P-matrix: It provides numerical validation for the existence and uniqueness of a solution to the LCP (q, M) .
- **[P-LCP]** Solve the LCP (q, M) given that $M \in \mathbf{P}$.
- **[P-LCP*]** Solve the LCP (q, M) or exhibit a nonpositive principal minor of M.

Facts and information:

- If P-LCP or P-LCP* is NP-hard, then NP = coNP [MP91].
- Comprehensive information on P-LCP can be found in the dissertation [Rus07].
- It is not known whether or not P-LCP can be achieved in polynomial time. It belongs to a complexity class with some other important problems, e.g., [TFNP] – problems for Total Functions from NP.

4.9.2 Univalence

Given a differential function $F\colon R \longrightarrow \mathbb{R}^n$, where R is a closed rectangle in $\mathbb{R}^n$, consider its Jacobian $F'(x)$. The Gale–Nikaido Theorem [GN65] states that if $F'(x) \in \mathbf{P}$ for all $x \in R$, then F is **univalent**, i.e., injective. An account of univalence theorems can be found in [Parth83]. It has also been shown that $F'(x) \in \mathbf{P}$ for merely all x on the boundary of R is sufficient to imply univalence of F [GZ79]. Subsequent work in [GR00] extended the connection between univalence and P-matrices to nonsmooth functions.

4.9.3 Interval Matrices

In the context of mathematical programming and elsewhere, the analysis of matrix intervals and interval equations play an important role and are intimately related to P-matrices. In particular, it is known that $[A,B]$ consists exclusively of n-by-n invertible matrices (i.e., $[A,B]$ is a **regular interval**) if and only if for each of the 2^n matrices $T \in [0,I]$ having diagonal entries in $\{0,1\}$, $TA+(I-T)B$ is invertible [Roh89]. Related work is presented in [Roh91]. The case of matrix intervals that consist exclusively of P-matrices is considered in [BG84] and [RR96] (see also [JR99]). It has been shown that $[A,B]$ consists entirely of P-matrices (i.e., $[A,B]$ is a **P-matrix interval**) if and only if for every nonzero $x \in \mathbb{R}^n$, there exists $j \in \langle n \rangle$ such that $x_j\,(Cx)_j > 0$ for every $C \in [A,B]$ (cf. Theorem 4.3.4).

4.9.4 Linear Differential Inclusion Problems

Consider the **Linear Differential Inclusion** (LDI) system

$$\dot{x} \in \Omega x, \quad x(0) = x_0, \quad \Omega \subset M_n(\mathbb{R})$$

and, in particular, the Diagonal Norm-bound LDI (DNLDI), which is a linear differential system with uncertain, time-varying (bounded) feedback gains. It leads to Ω being of the form

$$\Omega = \{A + B\Delta(I - E\Delta)^{-1}C\colon\ \|\Delta\| \le 1,\ \Delta = \text{diagonal}\}.$$

The following result is in [Gha90] (see also [BEFB94]).

Theorem 4.9.3 DNLDI is well posed if and only if $(I+E)^{-1}(I-E) \in \mathbf{P}$.

4.10 Miscellaneous Related Facts and References

We summarize and provide references to facts about P-matrices that are related to the themes of this chapter.

- There are several interesting generalizations of the notion of a P-matrix.
  - **P-tensors** are defined in [DLQ18] using an extension of the notion of non-reversal of sign of nonzero real vectors (cf. Theorem 4.3.4). They are shown to have properties and be related to the tensor positivity classes they generalize analogously to P-matrices.
  - In [KS14], the authors generalize the results in [JT95] by replacing the inverse appearing in Theorem 4.3.9 with the Moore–Penrose or the group inverse of the factor C.
  - In [SGR99], the notion of the **row-P-property** of a set of matrices is introduced and related to the unique solvability of nonlinear complementarity problems.
  - Generalizations of P-matrices to block forms based on extensions of Theorem 4.3.9 are undertaken and compared in [ES98]. Convex sets of P-matrices and **block P-matrices**, as well as (Schur) stability are considered in [ES00] and [EMS02].
  - P-matrix concepts and properties (like the non-reversal of the sign of a real nonzero vector, solutions to LCP-type problems and analogues to the relations between Z-matrices, M-matrices, P-matrices, and semipositivity) have been extended to transformations on **Euclidean Jordan Algebras**; see [GST04, TG05, GS06, GTR12].

- Sufficient conditions for positive stability of almost symmetric and **almost diagonally dominant P-matrices** are developed in [TSOA07].
- In [JK96], the authors consider matrices A some of whose entries are not specified and raise the problem of when do the unspecified entries can be chosen so that the completed matrix is in **P**. They show that if the specified entries include all of the diagonal entries, if every fully specified principal submatrix is in **P**, and if the unspecified entries are symmetrically placed (i.e., (i,j) entry is specified if and only if the (j,i) entry is specified), then the matrix A has a **completion to a P-matrix**. In [DH00], the authors extend the aforementioned class of partially specified matrices that have a P-matrix completion by providing strategies to achieve its conditions. In [DH00, JK96], partially specified matrices that cannot be completed to be in **P** are also discussed. General necessary and sufficient conditions for the completion of a partially specified matrix to a P-matrix are not known to date. Finally, completion problems of subclasses of **P** are studied in [FJTU00, GJSW84, Hog98a, Hog98b, JS96].
- In [BHJ85, HJ86a], the authors study linear transformations of **P** into **P**. In [BHJ85] it is shown that the linear transformations that map **P** *onto* itself are necessarily compositions of transposition, permutation similarity,

signature similarity, and positive diagonal equivalence (cf. Theorem 4.3.2 (1)–(4), respectively). Under the assumption that the kernel of a linear transformation $\mathcal{L}$ intersects trivially the set of matrices with zero diagonal entries, it is shown in [HJ86a] that the linear transformations that map **P** *into* itself ($n \geq 3$) are compositions of transposition, permutation similarity, signature similarity, positive diagonal equivalence, and the map $A \rightarrow A+D$, where D is a diagonal matrix whose diagonal entries are nonnegative linear combinations of the diagonal entries of A.

- There is an open problem of studying when the Hadamard (i.e., entry-wise) product of two P-matrices is a P-matrix. A related problem concerns Hadamard product of inverse M-matrices. In particular, it is conjectured that the Hadamard square of two inverse M-matrices is an inverse M-matrix; see [Neu98, WZZ00].
- The P-problem (detecting a P-matrix) is considered in [Rum03], where the sign-real spectral radius and interval matrix regularity are used to develop necessary and sufficient conditions for a real matrix to be in **P**. As a result, a not *a priori* exponential method for checking whether a matrix is in **P** or not is given.
- An algorithmic characterization of P-matrices via the Newton-min algorithm, which is an iterative method for the solution of the Linear Complementarity Problem, is presented in [GG19].
- Factorization of real matrices into products of P-matrices are considered in [JOvdD03]. In particular, it is shown that every $A \in M_n(\mathbb{R})$ can be written as the product of three P-matrices, provided that $\det A > 0$.
- The **Leading Implies All** (LIA) class of matrices comprises all real square matrices for which the positivity of the leading principal minors implies that all principal minors are positive. Thus LIA is contained in **P**. LIA matrices are studied in [JN13].
- Necessary conditions for a real matrix to be in **P** in terms of row and column sums are presented in [Szu90].
- Given a P-matrix $A \in M_n(\mathbb{R})$, the quantity

$$\alpha(A) = \min_{\|x\|_\infty} \{ \max_{1 \leq i \leq n} x_i(Ax)_i \}$$

and its role in the error analysis for the Linear Complementarity Problem is studied in [MP90] and [XZ02].

5

Inverse M-Matrices

5.1 Introduction

An n-by-n real matrix $A = [a_{ij}]$ is called an **M-matrix** ($A \in \mathbf{M}$) if

(1) it is of the form $A = \alpha I - B$, where B has nonnegative entries, and

(2) $\alpha > \rho(B)$ in which $\rho(B)$ is the **Perron–Frobenius eigenvalue** (spectral radius) of B.

Thus, an M-matrix A has two key features: the sign pattern $a_{ii} > 0$, $i = 1, \dots, n$ and $a_{ij} \le 0, i \ne j$ (such matrices are called **Z-matrices**, $A \in \mathbf{Z}$), and the property that all eigenvalues have positive real part (such matrices are called **positive stable**).

It is easily seen that an M-matrix A has a positive eigenvalue with minimum modulus, denoted by $q(A)$.

Equivalently, it can be shown that a Z-matrix A is an M-matrix if and only if it is invertible and $A^{-1} \ge 0$. Those matrices C that are inverses of M-matrices are called **inverse M-matrices** ($C \in \mathbf{IM}$) and comprise a large class of stable nonnegative matrices. Note that an inverse M-matrix is a nonnegative matrix whose inverse is a Z-matrix. A number of equivalent properties can be substituted for (2) above, such as those given in [Ple77, NP79, NP80]. We note that characterizations of inverse M-matrices are much harder to come by than those for M-matrices, and we present several that have come up in the literature.

Over the past half-century, M-matrices have had considerable attention, in large part because of the frequency with which they arise in applications [BP94], and a great deal is known about them and several generalizations, e.g., [FP62, NP79, NP80, Ple77, PB74]. M-matrices arise in iterative methods in numerical analysis of dynamical systems, finite difference methods for partial differential equations, input-output production and growth models in economics, linear complementarity problems in operations research, and

Markov processes in probability and statistics [BP94]. A significant amount of attention has focused on inverse M-matrices. Again, this is, in part, due to applications [Wil77]. In addition to inverse problems involving M-matrices, inverse M-matrices themselves arise in a number of applications such as the Ising model of ferromagnetism [Ple77], taxonomy [BP94], and random energy models in statistical physics [Fan60].

Notation We denote the n-by-n entry-wise nonnegative matrices by $\mathbf{N}$ (or that they are ≥ 0), and those with positive main diagonal by $\mathbf{N}^+$. The n-by-n diagonal matrices with positive diagonal entries are denoted by D_n^+. the n-by-n real matrices whose principal minors are positive by $\mathbf{P}$, the n-by-n matrices with nonpositive off-diagonal entries by $\mathbf{Z}$, the n-by-n M-matrices by $\mathbf{M}$, and the n-by-n inverse M-matrices by $\mathbf{IM}$.

Much is known about each of these important classes [And80, BP94, FP62, FP67, Joh78, Ple77, NP79, NP80]. If $A \in \mathbf{N}$, then $A \in \mathbf{IM}$ if and only if $A^{-1} \in \mathbf{Z}$, just as it is the case that if $B \in \mathbf{Z}$, then $B \in \mathbf{M}$ if and only if $B^{-1} \in \mathbf{N}$. Thus the classes $\mathbf{Z}$ and $\mathbf{N}$ are dual to each other in this context.

All inequalities between matrices or vectors of the same size, such as $A \geq B$, will be entry-wise. Furthermore, "$\geq$" gives an appropriate partial order for the classes $\mathbf{M}$ and $\mathbf{IM}$ as $A, B \in \mathbf{M}$ satisfy $B \geq A$ if and only if A^{-1}, $B^{-1} \in \mathbf{IM}$ satisfy $A^{-1} \geq B^{-1}$. Also, $A \in \mathbf{M}$ and $B \geq A$ imply $B \in \mathbf{M}$. (Both are straightforward calculations.) There is no corresponding fact for $\mathbf{IM}$, except Theorem 5.5.2; see also Theorem 5.17.23.

5.2 Preliminary Facts

A number of facts follow from the definition of M-matrices; many of these are presented without proof. Several equivalences for M-matrices are listed below for future use.

Theorem 5.2.1 [BP94] Let $A \in \mathbf{Z} \cap M_n(\mathbb{R})$. The following are equivalent.

(i) A is an M-matrix.
(ii) $A = \alpha I - B$ in which $B \geq 0$ and $\alpha > \rho(B)$.
(iii) A is invertible and $A^{-1} \geq 0$.
(iv) A has positive principal minors.
(v) There is $D \in D_n^+$ such that $DA + A^T D$ is positive definite, i.e., A has a **diagonal Lyapunov solution**.

Note that *(iv)* was proved in [FP62], and from this it follows that if $A \in \mathbf{IM}$, then $\det A > 0$ and A has positive diagonal entries.

5.2.1 Multiplicative Diagonal Closure

It is a familiar fact that the M-matrices are closed under positive diagonal multiplication; that is, if D if is diagonal with positive diagonal entries and $A \in \mathbf{M}$, then $DA \in \mathbf{M}$ and $AD \in \mathbf{M}$. The same is true of inverse M-matrices, for precisely the same reason.

Corollary 5.2.2 If $D \in D$ and $B \in \mathbf{IM}$, then $DB \in \mathbf{IM}$ and $BD \in \mathbf{IM}$.

Proof Because $DB \in \mathbf{N}$ and is invertible, it suffices to note that $(DB)^{-1} = B^{-1}D^{-1} \in \mathbf{Z}$, so that $(DB)^{-1} \in \mathbf{M}$ and $DB \in \mathbf{IM}$. The same argument applies to BD. □

5.2.2 Multiplicative Closure

Neither $\mathbf{M}$ nor $\mathbf{IM}$ is closed under multiplication.

Corollary 5.2.3 If $A \in \mathbf{N}$ is invertible, then $A \in \mathbf{IM}$ if and only if $A^{-1} \in \mathbf{Z}$.

Corollary 5.2.4 If $A, B \in \mathbf{IM}$, then $AB \in \mathbf{IM}$ if and only if $(AB)^{-1} \in \mathbf{Z}$.

Example 5.2.5 Consider the $\mathbf{IM}$ matrices

$$A = \begin{bmatrix} 21 & 14 & 13 \\ 13 & 24 & 20 \\ 14 & 13 & 24 \end{bmatrix}, \quad B = \begin{bmatrix} 24 & 20 & 13 \\ 13 & 24 & 14 \\ 14 & 13 & 24 \end{bmatrix}.$$

AB is not $\mathbf{IM}$ because the $(1, 3)$ entry of $(AB)^{-1}$ is positive.

Remark 5.2.6 Multiplicative closure can be shown to hold for $n = 2$ (because A, B, and AB have positive determinant).

5.2.3 Additive Closure

Another parallel between $\mathbf{IM}$ and $\mathbf{M}$ is that neither is a cone because neither is closed under addition. However, $\mathbf{M}$ is closed under addition in the 2-by-2 case.

Theorem 5.2.7 If $A, B \in \mathbf{IM}$, then $A + B \in \mathbf{IM}$ if and only if $A + B$ is invertible and $(A + B)^{-1} \in \mathbf{Z}$.

Additive closure does not even hold for $\mathbf{IM}$ matrices of order 2.

Example 5.2.8 Consider the $\mathbf{IM}$ matrix

$$A = \begin{bmatrix} 3 & 5 \\ 1 & 3 \end{bmatrix}.$$

If $B = A^T$, then $A + B$ is not even invertible. However, if $\det(A + B) > 0$, then $A + B$ is **IM** in the 2-by-2 case.

Remark 5.2.9 We note that, in case $n = 2$, for $A, B \in \mathbf{M}$, $A + B \in \mathbf{M}$ if and only if $A^{-1} + B^{-1} \in \mathbf{IM}$. This is not the case for $n > 2$, as indicated by the pair

$$A = \begin{bmatrix} 1 & 0 & 0 \\ -1 & 1 & 0 \\ -1 & -1 & 1 \end{bmatrix}, \quad B = \begin{bmatrix} 1 & 0 & 0 \\ 0 & 1 & -3 \\ 0 & 0 & 1 \end{bmatrix}$$

for which $A + B \in \mathbf{M}$, but $A^{-1} + B^{-1} \notin \mathbf{IM}$. There are, however, further interesting relationships between $A + B$ and $A^{-1} + B^{-1}$ to be noticed. For example, if $A, B \in \mathbf{P}$, $\det(AB) \cdot \det\left(A^{-1} + B^{-1}\right) = \det(A + B)$. Also, if A is lower and B is upper triangular, then the signs of the leading principal minors of $A^{-1} + B^{-1}$ and $A + B$ are the same, so that $A + B \in \mathbf{M}$ if $A^{-1} + B^{-1} \in \mathbf{P}$.

From Theorem 5.2.1 and the cofactor form of the inverse, we have

Theorem 5.2.10 If $A \in \mathbf{N}$, then $A \in \mathbf{IM}$ if and only if $\det A > 0$ and either $\det A(i,j) > 0$ or **sgn** $\det A(i,j) = (-1)^{i+j+1}$ for $-1 \leq i,j \leq n, i \neq j$.

IM is closed under permutation similarity and translation.

Theorem 5.2.11 If P is a permutation matrix, then $A \in \mathbf{IM}$ if and only if $PAP^T \in \mathbf{IM}$.

Theorem 5.2.12 $A \in \mathbf{IM}$ if and only if $A^T \in \mathbf{IM}$.

Our next fact is immediate from the corresponding property for M-matrices.

Theorem 5.2.13 Each matrix $A \in \mathbf{IM}$ is positive stable.

5.3 Diagonal Closure

Recall that $C = [c_{ij}]$ is said to be **diagonally dominant of its rows** if

$$|c_{ii}| > \sum_{j \neq i} |c_{ij}|, \quad i = 1, \ldots, n.$$

If C^T is diagonally dominant of its rows, then C is said to be **diagonally dominant of its columns**. The M-matrices possess an important (and characterizing) latent diagonal dominance property: if $A \in \mathbf{Z}$, then $A \in \mathbf{M}$ if and only if there is a diagonal D with nonnegative diagonal entries such that AD is diagonally dominant of its rows. Alternatively, DA may be made diagonally dominant of its columns, and, furthermore, DAE may be made diagonally dominant of both its rows and its columns. On the other hand, inverse M-matrices are not diagonally dominant of its rows (columns) in general. However, the inverse M-matrices do possess analogous dominance properties, which are dual (in a certain sense) rather than being exactly the same. The matrix C is said to be **diagonally dominant of its row entries** if

$$|c_{ii}| > |c_{ij}|,\ j \neq i,\ i = 1, \dots, n.$$

Similarly, C is said to be **diagonally dominant of its column entries** if C^T is diagonally dominant of its row entries. It is noted in [FP62, Joh77] and used in [Joh77, Wil77] that this weaker sort of dominance is latent in inverse M-matrices, just as the stronger sort is in M-matrices. We prove that fact here. Our approach is to scale the columns of $A \in \mathbf{M}$ so that the result is diagonally dominant of its rows, and then show, by inspection of the cofactors, that the inverse of the resulting matrix is diagonally dominant of its column entries.

Theorem 5.3.1 [BP94] If $A \in \mathbf{IM} \cap M_n(\mathbb{R})$, then the following are equivalent.

(i) There is a $D \in D_n^+$ such that DA is diagonally dominant of its column entries.
(ii) There is an $E \in D_n^+$ such that AE is diagonally dominant of its row entries.
(iii) There are $G, H \in D_n^+$ such that GAH is diagonally dominant of both its row and column entries.

Proof Let $A \in \mathbf{IM}$ so that $B = A^{-1} \in \mathbf{M}$. It follows from the Perron–Frobenius Theorem [HJ13] that there is a $D \in D_n^+$ such that $R = BD$ is diagonally dominant of its rows. Let $S = R^{-1} = D^{-1}A = [s_{ij}]$ and suppose $i, j \in \langle n \rangle$ with $i \neq j$. Then it follows from the cofactor expansion of R^{-1} and from $R \in \mathbf{Z}$ that

$$|s_{ii}| - |s_{ji}| = \frac{\det R[\langle n \rangle \setminus \{i\}] + \det[R[\langle n \rangle \setminus \{i\}, \langle n \rangle \setminus \{j\}]}{\det R} = \frac{\det T}{\det R}$$

in which T is obtained from $R[\langle n \rangle \setminus \{i\}]$ by adding the column $\pm R[\langle n \rangle \setminus \{i\}, i]$ to the j-th column. Because R is diagonally dominant of its rows and has positive diagonal, $\det T > 0$, and so $|s_{ii}| - |s_{ji}| > 0$. Thus, $S = D^{-1}A$ is diagonally dominant of its column entries and (i) holds. The proof of (ii) is similar, and (iii) follows from (i) and (ii). □

Of course, because of Theorem 5.3.1, any **IM** matrix may be diagonally scaled to one with 1s on the diagonal (thereby obtaining a **normalized inverse M-matrix**). In fact, the scaled **IM** matrix may be taken to have 1s on the off-diagonal and entries < 1 [JS01a].

5.4 Diagonal Lyapunov Solutions

It is known that M-matrices have **diagonal Lyapunov solutions**; that is, for each $B \in \mathbf{M}$, there is a $D \in D_n^+$ such that $DB + B^T D$ is positive definite. (This follows from the fact that an M-matrix may be scaled to have both row and column diagonal dominance [JS01a].) This allows us to prove the analogous fact for **IM** matrices.

Theorem 5.4.1 For each $A \in \mathbf{IM} \cap M_n(\mathbb{R})$, there is a $D \in D_n^+$ such that $DA + A^T D$ is positive definite.

Proof $A = B^{-1}$ for some $B \in \mathbf{M}$. Let $D \in D_n^+$ such that $DB + B^T D$ is positive definite. Equivalently, $DA^{-1} + \left(A^{-1}\right)^T D$ is positive definite. Hence, $A^T \left(DA^{-1} + \left(A^{-1}\right)^T D\right) A = DA + A^T D$ is also positive definite. □

It is worth noting, moreover, that the set of all possible diagonal Lyapunov solutions of A (which is a cone) is the same as that for $B = A^{-1}$. Because principal submatrices of positive definite matrices are positive definite, it follows that the possession of a diagonal Lyapunov solution is a property inherited under the extraction of principal submatrices. Because the possession of a positive definite Lyapunov solution implies positive stability, which implies a positive determinant for a real matrix, it means that an M-matrix is a P-matrix, a familiar fact. The same argument applies to Theorem 5.4.1.

Corollary 5.4.2 The class $\mathbf{IM} \subseteq \mathbf{P}$.

We note that $\mathbf{Z} \cap \mathbf{P} = \mathbf{M}$, but $\mathbf{N} \cap \mathbf{P} \neq \mathbf{IM}$. Within the class **Z**, the existence of a diagonal Lyapunov solution characterizes **M** (Theorem 5.2.1). Theorem 5.4.1 may be interpreted as saying that the existence of a diagonal Lyapunov solution is necessary for $A \in \mathbf{N}$ to be in **IM**. It is not a sufficient condition, however.

Example 5.4.3

$$A = \begin{bmatrix} 18 & 1 & 6 \\ 1 & 18 & 6 \\ 6 & 6 & 18 \end{bmatrix}$$

has the identity as a Lyapunov solution, but its inverse is not in **Z**.

It should be noted that, for $n = 2$, a matrix $A \in \mathbf{N}$ is in **IM** if and only if $\det A > 0$ (just as $B \in \mathbf{Z}$ is in **M** if and only if $\det B > 0$ for $n = 2$), but this easy characterization is atypical.

5.5 Additive Diagonal Closure

Although it was easy to see the closure of **IM** under positive diagonal multiplication from the corresponding fact for **M**, it is somewhat more subtle that the parallel extends to addition.

It is known that for $A \in \mathbf{M} \cap M_n(\mathbb{R})$ and $D \in D_n^+$, $A + D \in \mathbf{M}$.

This fact may be seen, for example, from diagonal Lyapunov solution, from diagonal-dominance characterizations, or from the **von Neumann expansion**

$$(I - A)^{-1} = I + A + A^2 + \cdots$$

if $\rho(A) < 1$. The latter is given in [JS01a]. Here, we provide a proof based upon diagonal dominance characterizations.

Theorem 5.5.1 If $A \in \mathbf{IM}$ and E_{ii} is the n-by-n matrix with a 1 in the (i, i) position and 0s elsewhere, then $A + tE_{ii} \in \mathbf{IM}$ for any $t \geq 0$.

Proof Let $A \in \mathbf{IM}$ and let $A^{-1} = [\alpha_{ij}]$. Without loss of generality, we may assume that $i = 1$. Let $B = A + te_1e_1^T$ in which e_1 denotes the first standard basis vector. Then, from [HJ13, p. 19], we have

$$\begin{aligned} B^{-1} &= A^{-1} - \frac{1}{1 + e_1^T A^{-1} e_1} A^{-1} e_1 e_1^T A^{-1} \\ &= A^{-1} - \frac{1}{1 + \alpha_{11}} \begin{bmatrix} \alpha_{11} \\ \alpha_{21} \\ \vdots \\ \alpha_{n1} \end{bmatrix} [\alpha_{11}\ \alpha_{12}\ \cdots\ \alpha_{1n}] \\ &= A^{-1} + \begin{bmatrix} - & + & + & . & . & + \\ + & - & - & . & . & - \\ + & - & - & . & . & - \\ . & . & . & . & . & . \\ . & . & . & . & . & . \\ + & - & - & . & . & - \end{bmatrix} \\ &= A^{-1} + C. \end{aligned}$$

For $j \neq 1$, $|\alpha_{1j}| = \frac{\alpha_{11}}{1+\alpha_{11}} |\alpha_{1j}| = c_{1j}$ and $|\alpha_{j1}| \geq \frac{\alpha_{11}}{1+\alpha_{11}} |\alpha_{j1}| = c_{j1}$. Thus, $B^{-1} \in \mathbf{Z}$, which implies $B \in \mathbf{Z}$ and completes the proof. □

Closure under addition of a nonnegative diagonal matrix follows immediately.

Theorem 5.5.2 If $A \in \mathbf{IM} \cap M_n(\mathbb{R})$ and $D \in D_n^+$, then $A + D \in \mathbf{IM}$.

This observation may also be made via an inductive argument based upon a characterization, to be given later, of how an inverse M-matrix may be embedded in another of larger dimension.

Remark 5.5.3 We note that only the matrices in D_n^+ have the above property; i.e., if E is such that $A+E \in \mathbf{IM}$ for all $A \in \mathbf{IM}$, then E must be a nonnegative diagonal matrix. This is straightforward, and we leave details to the reader.

Because Theorem 5.5.2 indicates that we may increase diagonal entries and stay in **IM**, and Theorem 5.3.1 indicates that, in some sense, off-diagonal entries must be smaller than diagonal entries, a natural question to address is whether we may decrease off-diagonal entries and remain **IM**. This is essentially the case for **M** because of the diagonal-dominance characterization. That is, if any off-diagonal entry of $A \in \mathbf{M}$ is decreased in absolute value (increased algebraically, but not past 0), the result is still in **M**. Unfortunately, this is not the case for **IM** as illustrated by A and B in the next example, which differ only in the $(1,3)$ entry.

Example 5.5.4

$$A = \begin{bmatrix} 4 & 2 & 1 \\ 1 & 4 & 2 \\ 2 & 1 & 4 \end{bmatrix} \in \mathbf{IM} \quad \text{and} \quad B = \begin{bmatrix} 4 & 2 & 0 \\ 1 & 4 & 2 \\ 2 & 1 & 4 \end{bmatrix} \notin \mathbf{IM}.$$

If, however, all the off-diagonal entries in some row (or column) are decreased by a common factor of scale, the outcome is different.

Theorem 5.5.5 Suppose that $A = [a_{ij}] \in \mathbf{IM}$ and that $A_k(\theta) = [a_{ij}(\theta)]$ is defined by

$$a_{ij}(\theta) = \begin{cases} \theta a_{ij} & \text{for } i = k \text{ and } j \neq k, \\ a_{ij} & \text{otherwise.} \end{cases}$$

Then $A_k(\theta) \in \mathbf{IM}$ for $0 \leq \theta \leq 1$ and $1 \leq k \leq n$.

Proof Write $A_k(\theta) = DA + E$ where D is the diagonal matrix agreeing with I except for a θ in the k-th diagonal position, and E is the zero matrix except for $(1-\theta)a_{kk}$ in the k-th diagonal position. Then, for $k > 0$, $A_k(\theta) \in \mathbf{IM}$ by application of Corollary 5.2.2 and Theorem 5.5.2. The case of $\theta = 0$ may be handled by taking limits. □

An implication of Theorem 5.5.5 is worth noting.

Corollary 5.5.6 If $A \in \mathbf{IM}$ and B is a principal submatrix of A, then $B \in \mathbf{IM}$.

Proof Multiple applications (to A and A^T) of Theorem 5.5.5 for $\theta = 0$ leave a matrix in **IM** that is permutation similar to the direct sum of B and D for some D in $\mathcal{D}$. Its inverse is in **M** and is permutation similar to the direct sum of B^{-1} and D^{-1}. Because $B^{-1} \in \mathbf{Z}$, it follows that $B \in \mathbf{IM}$. □

Closure of **IM** under submatrix extraction was originally demonstrated in [Mark72] by very different means and parallels the corresponding fact for **M**. Because (1) principal minors are determinants of principal submatrices, and (2) matrices in **M**and **IM** have positive determinants, we note that Corollary 5.5.6 implies Corollary 5.4.2.

5.6 Power Invariant Zero Pattern

Another property of inverse M-matrices has been noted (and even generalized) by several authors [LN80, SRPM82] and is of interest when some off-diagonal entries are 0. That is, the zero-nonzero patterns of inverse M-matrices are power invariant. (This is in contrast with the fact that there is no special restriction on off-diagonal zero entries of M-matrices.) We state this result, which has been proven elsewhere.

Theorem 5.6.1 Suppose that $A \in \mathbf{IM}$, and let k be a positive integer. Then, the (i,j) entry of A^k is zero if and only if the (i,j) entry of A is zero.

We know from Theorem 5.2.1 that $\mathbf{IM} \subseteq \mathbf{N}^+$. In $\mathbf{N}^+$, the property of having a power invariant zero pattern is purely combinatorial and, in fact, is equivalent to A and A^2 having the same zero pattern. Furthermore, it should be noted that (1) the values of the diagonal entries are immaterial to the question of a power invariant zero pattern and (2) if there are zeros, a power invariant zero pattern must be reducible, and each irreducible component must be strictly positive. In view of Theorems 5.5.2 and 5.3.1, a natural question to raise is: for which $A \in \mathbf{N}^+$ (or, equivalently, **N**) does there exist a diagonal D with nonnegative diagonal entries such that $A + D \in \mathbf{IM}$? Theorem 5.6.1 gives a necessary condition and shows that not all $A \in \mathbf{N}^+$ may be put in **IM** by increasing the diagonal. However, this power invariance of the zero pattern is the only restriction, and except for it, the diagonal is the crucial feature associated with membership in **IM**, just as it is for **M**. Thus, the question is purely combinatorial, and its answer, by appealing to the von Neumann expansion, generalizes Theorem 5.6.1.

Theorem 5.6.2 Suppose that $A \in \mathbf{N}^+ \cap M_n(\mathbb{R})$. Then there is a $D \in D_n^+$ such that $A + D \in \mathbf{IM}$ if and only if A^2 and A have the same zero pattern.

Proof If $A + D \in \mathbf{IM}$ for some $D \in D_n^+$, then the necessity of the same zero pattern in A^2 and A follows from Theorem 5.6.1 (which may also be deduced from the calculations to be given below). On the other hand, if A and A^2 have the same zero pattern, then the pattern of A is power invariant, and it suffices to show that $\alpha \geq 0$ may be chosen so that $(\alpha I + A)^{-1}$ exists and is in $\mathbf{Z}$. First, suppose that $(\alpha I + A)^{-1}$ exists and is in $\mathbf{Z}$. Next, suppose that $\alpha > \rho(A)$. Then,

$$(\alpha I + A)^{-1} = \frac{1}{\alpha}\left(I - \frac{1}{\alpha}A + \frac{1}{\alpha^2}A^2 - \frac{1}{\alpha^3}A^3 + \cdots\right).$$

It then suffices to show that α can be chosen so that

$$\operatorname{diag}\left(\frac{1}{\alpha}A - \frac{1}{\alpha^2}A^2 + \frac{1}{\alpha^3}A^3 - \cdots\right) < I \tag{i}$$

and

$$\left(A - \frac{1}{\alpha}A^2 + \frac{1}{\alpha^2}A^3 - \cdots\right) \geq 0. \tag{ii}$$

For then we would have $(\alpha I + A)^{-1} \in \mathbf{Z}$. Note that (i) is equivalent to

$$\operatorname{diag}\left[A(\alpha I + A)^{-1}\right] < I, \tag{i$'$}$$

and (ii) is equivalent to

$$A^2(\alpha I + A)^{-1} \leq A. \tag{ii$'$}$$

Because $(\alpha I + A)^{-1} \to 0$ as $\alpha \to \pm\infty$, it is clear that (i′) is satisfied for sufficiently large α. Furthermore, because A has power invariant zero pattern, $A^2(\alpha I + A)^{-1}$ and A have the same zero pattern, so that the fact that the left-hand side of (ii′) can be made arbitrarily small by the choice of $\alpha > 0$ means that (ii′) may be satisfied. Because the convergence of $(\alpha I + A)^{-1}$ and (i) and (ii) may all be satisfied for α sufficiently large, they may all be satisfied simultaneously, which completes the proof. □

Question 5.6.3 [Joh82] Consider $A \in \mathbf{N}$ such that $\alpha I + A$ has power invariant zero pattern for $\alpha > 0$. According to Theorem 5.5.2 and Theorem 5.6.2, there exists a real number α_0 such that $\alpha I + A \in \mathbf{IM}$ for $\alpha > \alpha_0$ and $\alpha I + A \notin \mathbf{IM}$ for $\alpha \leq \alpha_0$. How may α_0, a function of A, be characterized? The corresponding M-matrix question has a pleasantly simple answer (namely, if and only if $\alpha > \rho(A)$) with no combinatorial requirement.

The combinatorial portion of the answer is straightforward in the case of $\mathbf{IM}$ (Theorem 5.6.2); however, the analytical portion (determination of $\alpha_0(A)$) does not appear to be as simple (and not as neatly related to the spectrum). If an answer were available, a characterization of $A \in \mathbf{N}^+$ that lie in $\mathbf{IM}$ would follow: for example, $D \in \mathbf{IM}$ could be chosen so that $\text{diag}(DA) = I$ and $P = DA - I$; then $\alpha_0(P)$ could be compared to 1 to provide an answer (along with a check of power invariance). This question was answered in [JS07a, theorem 4.16.2].

Question 5.6.4 [Joh82] Because $A, B \in \mathbf{Z}$, $B - A \in \mathbf{N}$, and $A \in \mathbf{M}$ imply $B \in \mathbf{M}$, it seems natural to ask, more generally, under what circumstances are $A, B \in \mathbf{N}$, $B - A \in \mathbf{N}$, and $A \in \mathbf{IM}$ imply $B \in \mathbf{IM}$? Theorems 5.5.2, 5.5.5 (and the discussion preceding), and 5.6.2 give special cases, but there may be a more encompassing answer. In general, we might ask: given $A \in \mathbf{N}$, which satisfies some condition necessary for $A \in \mathbf{IM}$, how might the entries of A be adjusted so that the resulting matrix lies in $\mathbf{IM}$?

5.7 Sufficient Conditions for a Positive Matrix to Be IM

Thus far, the conditions we have discussed have primarily been only necessary conditions for $A \in \mathbf{IM}$. One paper, which also contains several other facts about $\mathbf{IM}$, primarily concentrates on developing a very special condition sufficient for a positive matrix to be an inverse M-matrix [Wil77]. Suppose now that a positive matrix $A = [a_{ij}]$ is diagonally scaled so that $a_{ii} = 1, i = 1, \dots, n$, and $a_{ij} < 1, i \neq j$. We know that any inverse M-matrix may be scaled this way and that the original matrix is $\mathbf{IM}$ if and only if A is. (Such scalings, it should be noted, are not unique because alternate ones may be obtained by diagonal similarity.) Further, let

$$a_{\max} = \max_{i \neq j} a_{ij}, \quad a_{\min} = \min_{i \neq j} a_{ij},$$

and define t by $a_{max}^2 = t a_{min} + (1 - t) a_{min}^2$. Under the stated conditions on A, we have $0 < a_{min} \leq a_{max} < 1$.

We state but do not re-prove the main result of [Wil77] with t as defined above.

Theorem 5.7.1 Let $A = [a_{ij}] > 0$ be such that $a_{ii} = 1, i = 1, \dots, n$, and $a_{ij} < 1, i \neq j$. Then $A \in \mathbf{IM}$ if $t^{-1} \geq n - 2$. Furthermore, the stated condition is tight; if $a_{ij} = a_{min}$, $i \neq j$, except for $a_{ij} = a_{ji} = a_{max}$ when $i = n - 1, n$ and $1 \leq j \leq n - 2$, then $A \in \mathbf{IM}$ if and only if $t^{-1} \geq n - 2$.

At the opposite extreme from considering when a positive A is $\mathbf{IM}$ is the interesting question: which triangular $A \in \mathbf{N}^+$ are inverse M-matrices? Of course, a triangular matrix in $\mathbf{Z}$ must be an M-matrix, but this is now another point of difference between $\mathbf{M}$ and $\mathbf{IM}$. If $A \in \mathbf{N}^+$ is triangular, then A need not be in **IM**; it still depends on the minors, which may have either sign. One extreme case to consider is which (0, 1) matrices are in **IM**. It is easy to see that such a matrix must have 1s on the diagonal and be permutation similar to a triangular matrix, but this still is not enough to characterize such matrices. In [LN80] the (0, 1) inverse M-matrices are characterized, and the presentation of the characterization is graph-theoretic.

Another interesting feature of triangular matrices in either **M** or **IM**, and another analogy between the two classes, is noted in [Mark79]. If some power of $A \in$ **IM**($B \in$ **M**) is permutation similar to a triangular matrix, then A (B) must be also (and by the same permutation). Necessary and sufficient conditions for an invertible, triangular, normalized **SPP** matrix (see Section 5.12) to be **IM** are given in Theorem 5.12.6. Factorizations LDU for inverse M-matrices have been studied. There the central question, which at first glance appears harder but is conceivably easier, is: which $A \in \mathbf{N}^+$ are (finite) products of inverse M-matrices? Clearly, any product of inverse M-matrices is in $\mathbf{N}^+$, but this does not cover $\mathbf{N}^+$, which leaves an intriguing question.

The special case of inverse M-matrices that are inverses of tridiagonal matrices is treated in [Lew80].

5.8 Inverse M-Matrices Have Roots in IM

In general, matrices in $\mathbf{N}^+$ do not necessarily have square roots (or higher-order roots) in $\mathbf{N}^+$ (or even **N**), although any power of a matrix in $\mathbf{N}^+$ is in $\mathbf{N}^+$. However, matrices in **IM** have arbitrary roots not only in $\mathbf{N}^+$ but in **IM**. As we have often seen, this is parallel to a corresponding fact for M-matrices.

Theorem 5.8.1 If $B \in \mathbf{M}$, then, for each integer $k \geq 1$, there is a matrix $B^{1/k}$ such that $(B^{1/k})^k = B$ and $\left(B^{1/k}\right)^q \in \mathbf{M},\ q = 0, 1, \ldots, k$.

Proof We may write $B = \alpha I - P$ where $P \geq 0$ and $\alpha > \rho(P)$. Thus, $(1/\alpha)B = I - (1/\alpha)P$, and, because it suffices to prove the theorem for $(1/\alpha)P$, we assume, without loss of generality, that $\alpha = 1$. For a scalar t, we have

$$(1 - x)^t = \sum_{i=0}^{\infty} (-1)^i \begin{bmatrix} t \\ i \end{bmatrix} x^i$$

for $|x| < 1$. Because $\rho(P) < 1$, it follows that

$$(I - P)^t = \sum_{i=0}^{\infty} (-1)^i \begin{bmatrix} t \\ i \end{bmatrix} P^i.$$

Because

$$\begin{bmatrix} t \\ i \end{bmatrix} = t(t-1)\ldots(t-i+1)/i!,$$

we have, for $0 < t < 1$, such that

$$(-1)^i \begin{bmatrix} t \\ i \end{bmatrix} < 0, i = 1, 2, \ldots,$$

and $-\sum_{i=1}^{\infty}(-1)^i \binom{t}{i} P^i$ converges to a nonnegative matrix $f(P)$, whose spectral radius $1 - [1 - \rho(P)]^T$ is less than 1. Thus, for $0 \leq t \leq 1$, $(I - P)^T = 1 - f(P) \in \mathbf{M}$. Setting $t = 1/k$ yields the asserted result. □

Corollary 5.8.2 If $\mathcal{A} \in \mathbf{IM}$, then, for each integer $k \geq 1$, there is a matrix $A^{1/k}$ satisfying $\left(A^{1/k}\right)^k = A$ and $\left(A^{1/k}\right)^q \in \mathbf{IM}$, $q = 0, 1, \ldots, k$.

Proof Let $A = B^{-1}$, $B \in \mathbf{M}$, and $A^{1/k} = \left(B^{1/k}\right)^{-1}$, where $B^{1/k}$ is the matrix given by Theorem 5.8.1. Thus, matrices in **M**(**IM**) have arbitrary k-th roots within the class **IM** (**M**). □

We note that, because they are high-order powers of matrices in $\mathbf{N}^+$, this is another explanation of why elements of **IM** have power invariant 0-patterns.

Question 5.8.3 [Joh82] Besides powers of elements of **IM**, are there any other invertible elements of $\mathbf{N}^+$ that, for each $k = 1, 2, \ldots,$ have k-th roots in $\mathbf{N}^+$? (Note that although **IM** is closed under the extraction of particular roots, powers of **IM** matrices are not necessarily **IM** because powers of M-matrices are not necessarily **M**.) However, a power of an element of **IM** does lie in $\mathbf{N}^+$and has arbitrary roots in $\mathbf{N}^+$because of Corollary 5.8.2.) If not, then **IM** would be characterized by the fact that powers of its elements have arbitrary roots in $\mathbf{N}^+$. A corresponding question is the following: must each element of $\mathbf{N}^+$ which has arbitrary roots in $\mathbf{N}^+$ have a root in **IM**?

Question 5.8.4 [Joh82] The invertible elements of $\mathbf{N}^+$which satisfy Hadamard's inequality form a semigroup. What is the structure of this semigroup, and how does it compare with the semigroup of products of elements of **IM**?

5.9 Partitioned IM Matrices

Summary. Throughout the rest of this chapter we assume that all relevant inverses exist. By considering necessary and/or sufficient conditions for

$A \in \mathbf{IM}$ when A is written in partitioned form, a great deal can be learned. For instance, Sylvester's matrix of "bordered" minors can be defined in terms of Schur complements, and the conformal form for A^{-1} can be defined in terms of inverses of matrices in $\mathbf{N}^{+}$. The partitioned form of A^{-1} together with the fact that M-matrices are closed under extraction of principal submatrices and Schur complementation allow us to characterize **IM** matrices and to determine a number of inequalities involving partitions of **IM** matrices. Schur's formula is derived as well as a special case of Sylvester's identity for determinants; these, in turn, are used to provide necessary and sufficient conditions for a nonnegative matrix to be **IM** (see Theorems 5.9.7, 5.9.8, 5.9.9). It is noted that necessary and sufficient conditions for a matrix to be **IM** are generally much harder to come by than those for M-matrices.

5.9.1 Sylvester's Matrix of Bordered Minors

If A is square and $A[\alpha]$ is invertible, recall that the Schur complement of $A[\alpha]$ in A, denoted $A/A[\alpha]$, is defined by

$$A/A[\alpha] = A[\alpha^c] - A[\alpha^c,\alpha]A[\alpha]^{-1}A[\alpha,\alpha^c].$$

Let $\alpha = \langle k\rangle$. (Due to Theorem 5.2.11, there is no difference between $\alpha = \langle k\rangle$ and a general α.) It was shown in [CH69] that if $A/A[\alpha] = B = [b_{ij}]$, then, for $k+1 \le i,j \le n$,

$$b_{ij} = \frac{\det A[\alpha + i, \alpha + j]}{\det A[\alpha]} = \frac{s_{ij}}{\det A[\alpha]},$$

in which $S = [s_{ij}]$ is Sylvester's matrix of "bordered" minors [Gan59], i.e.,

$$s_{ij} = \det \begin{bmatrix} A[\alpha] & A[\alpha, j] \\ A[i,\alpha] & a_{ij} \end{bmatrix}.$$

Thus, $S = (\det A[\alpha])(A/A[\alpha])$.

5.9.2 Schur Complement Form of the Inverse

We shall use the Schur complement form of the inverse [HJ13] given in the following form. Let the square matrix A be partitioned as

$$\mathbf{A} = \begin{bmatrix} A[\alpha] & A[\alpha,\alpha^c] \\ A[\alpha^c,\alpha] & A[\alpha^c] \end{bmatrix} \qquad (5.9.1)$$

in which A, $A[\alpha]$, and $A[\alpha^c]$ are all invertible. Then

$$\mathbf{A}^{-1} = \begin{bmatrix} (A/A[\alpha^c])^{-1} & -A[\alpha]^{-1}A[\alpha,\alpha^c](A/A[\alpha])^{-1} \\ -(A/A[\alpha])^{-1}A[\alpha^c,\alpha]A[\alpha]^{-1} & (A/A[\alpha])^{-1} \end{bmatrix}$$
$$= \begin{bmatrix} (A/A[\alpha^c])^{-1} & -(A/A[\alpha^c])^{-1}A[\alpha,\alpha^c]A[\alpha^c]^{-1} \\ -A[\alpha^c]^{-1}A[\alpha^c,\alpha](A/A[\alpha^c])^{-1} & (A/A[\alpha])^{-1} \end{bmatrix}. \tag{5.9.2}$$

We now make use of the fact that M-matrices are closed under extraction of principal submatrices and under extraction of Schur complements (Schur complementation) [HJ13, HJ91].

Theorem 5.9.1 Let $A \geq 0$ be partitioned as $A = \begin{bmatrix} A_{11} & A_{12} \\ A_{21} & A_{22} \end{bmatrix}$ in which A_{11} and A_{22} are non-void principal submatrices of A. Then, $A \in \mathbf{IM}$ if and only if

(i) $A/A_{11} \in \mathbf{IM}$;
(ii) $A/A_{22} \in \mathbf{IM}$;
(iii) $(A_{11})^{-1}A_{12}(A/A_{11})^{-1} \geq 0$;
(iv) $(A/A_{11})^{-1}A_{21}(A_{11})^{-1} \geq 0$;
(v) $(A_{22})^{-1}A_{21}(A/A_{22})^{-1} \geq 0$;
(vi) $(A/A_{22})^{-1}A_{12}(A_{22})^{-1} \geq 0$.

Proof For necessity, suppose $A \in \mathbf{IM}$, and consider the Schur complement form of its inverse. Because $A^{-1} \in \mathbf{M}$ and M-matrices are closed under extraction of principal submatrices, $(A/A_{11})^{-1}$ and $(A/A_{22})^{-1}$ are in $\mathbf{M}$, and (i) and (ii) follow. Statements (iii)–(vi) follow because $A^{-1} \in \mathbf{Z}$.

For sufficiency, observe that (i) and (ii) and either (iii) and (iv) or (v) and (vi) ensure that $A^{-1} \in \mathbf{Z}$. This completes the proof. □

Corollary 5.9.2 **IM** matrices are closed under extraction of Schur complements.

Corollary 5.9.3 **IM** matrices are closed under extraction of principal submatrices.

These follow from Theorem 5.9.1 and the Schur complement form of $(A^{-1})^{-1}$, respectively. In turn, Theorem 5.9.5 implies that

Corollary 5.9.4 **IM** matrices have positive principal minors.

Notice also that Theorem 5.9.1 allows us to zero out any row or column of an **IM** matrix off the diagonal and remain **IM**. That is, if

$$A(t) = \begin{bmatrix} a_{11} & A_{12} \\ tA_{21} & A_{22} \end{bmatrix}, \quad A = \begin{bmatrix} a_{11} & A_{12} \\ A_{21} & A_{22} \end{bmatrix} \in \mathbf{IM},$$

and $B = \begin{bmatrix} a_{11} & A_{12} \\ 0 & A_{22} \end{bmatrix}$, then

$$B^{-1} = \begin{bmatrix} (a_{11})^{-1} & -(a_{11})^{-1}A_{12}(A_{22})^{-1} \\ 0 & (A_{22})^{-1} \end{bmatrix} \in \mathbf{Z}$$

because $a_{11}, A_{12}(A_{22})^{-1} \geq 0$. This fact can also be shown by applying Corollary 5.2.2 and Theorem 5.5.1, i.e., multiply the first column of A by some t, $0 < t < 1$, then add $a_{11} - ta_{11} > 0$ to the $(1,1)$ entry to obtain the **IM** matrix $A(t) = \begin{bmatrix} a_{11} & A_{12} \\ tA_{21} & A_{22} \end{bmatrix}$. By continuity,

$$\begin{bmatrix} a_{11} & A_{12} \\ 0 & A_{22} \end{bmatrix}^{-1} = \begin{bmatrix} (a_{11})^{-1} & -(a_{11})^{-1}A_{12}(A_{22})^{-1} \\ 0 & (A_{22})^{-1} \end{bmatrix} \in \mathbf{IM}.$$

Matrix $B \in \mathbf{M}$ if and only if $B/B_{11}, B/B_{22} \in \mathbf{M}$, but this does not hold for **IM** matrices. (When partitioned as in Theorem 5.9.1, the $\mathbf{N}^+$matrix

$$A = \begin{bmatrix} 1 & 1 & 1 \\ 1 & 3 & 4 \\ 1 & 2 & 3 \end{bmatrix}$$

in which $A_{11} = a_{11}$ satisfies $A_{11}, A_{22}, A/A_{11}, A/A_{22}$ are **IM** provides a counterexample [Ima84] because $A \notin \mathbf{IM}$. Additional conditions are given so that $A/A_{11}, A/A_{22} \in \mathbf{IM}$ implies that $A \in \mathbf{IM}$.)

We also have

Theorem 5.9.5 Let $A \geq 0$ be partitioned as $A = \begin{bmatrix} A_{11} & A_{12} \\ A_{21} & A_{22} \end{bmatrix}$ in which A_{11} and A_{22} are non-void principal submatrices of A. Then, if $A \in \mathbf{IM}$,

(i) $A_{11} \in \mathbf{IM}$;
(ii) $A/A_{11} \in \mathbf{IM}$;
(iii) $A_{22} \in \mathbf{IM}$;
(iv) $A/A_{22} \in \mathbf{IM}$;
(v) $(A_{11})^{-1}A_{12} \geq 0$;
(vi) $A_{21}(A_{11})^{-1} \geq 0$;
(vii) $(A_{22})^{-1}A_{21} \geq 0$;
(viii) $A_{12}(A_{22})^{-1} \geq 0$;
(ix) $A_{12}(A/A_{11})^{-1} \geq 0$;
(x) $(A/A_{11})^{-1}A_{21} \geq 0$;
(xi) $A_{21}(A/A_{22})^{-1} \geq 0$;
(xii) $(A/A_{22})^{-1}A_{12} \geq 0$.

Proof Observe that (i)–(iv) follow from the preceding remarks. Then, (v)–(xii) follow from (iii)–(vi) of Thcorcm 5.9.1 upon multiplying by thc appropriate choice of A/A_{11}, A/A_{22}, A_{11}, or A_{22}, which completes the proof. □

For **IM** matrices of order 2 or 3, it is obvious from the remarks preceding Theorem 5.9.5 that we can zero out any **reducing block** (given a block matrix $\mathbf{A} = \left[\begin{smallmatrix} B & C \\ D & E \end{smallmatrix}\right]$, the block C, respectively, D, is a reducing block provided A, B, E are square) or zero out either triangular part and remain **IM**. However, these properties do not hold in general.

Example 5.9.6 Consider the **IM** matrix

$$\mathbf{A} = \begin{bmatrix} 20 & 8 & 11 & 5 \\ 19 & 20 & 19 & 12 \\ 17 & 8 & 20 & 5 \\ 14 & 9 & 14 & 20 \end{bmatrix}.$$

Neither $\mathbf{B} = \left[\begin{smallmatrix} 20 & 8 & 0 & 0 \\ 19 & 20 & 0 & 0 \\ 17 & 8 & 20 & 5 \\ 14 & 9 & 14 & 20 \end{smallmatrix}\right]$ nor $\mathbf{C} = \left[\begin{smallmatrix} 20 & 0 & 0 & 0 \\ 19 & 20 & 0 & 0 \\ 17 & 8 & 20 & 0 \\ 14 & 9 & 14 & 20 \end{smallmatrix}\right]$ is in **IM** because the $(4, 1)$ entry of the inverse of each is positive.

For other characterizations of **IM** matrices, we will utilize Schur's formula [HJ13], which states that

$$\det A = (\det A[\alpha])(\det A/A[\alpha])$$

provided $A[\alpha]$ is invertible. We will also apply a special case of Sylvester's identity for determinants, which follows.

Let A be an n-by-n matrix, $\alpha \subseteq \mathcal{N}$, and suppose $|\alpha| = k$. Define the $(n-k)$-by-$(n-k)$ matrix $B = [b_{ij}]$ by setting $b_{ij} = \det A[\alpha + i, \alpha + j]$, for every $i, j \in \alpha^c$. Then Sylvester's identity for determinants (see [HJ13]) states that for each $\delta, \gamma \subseteq \alpha^c$, with $|\delta| = |\gamma| = m$,

$$\det B[\delta, \gamma] = (\det A[\alpha])^{m-1} \det A[\alpha \cup \delta, \alpha \cup \gamma]. \tag{5.9.3}$$

Then, let A be an n-by-n matrix partitioned as follows:

$$\mathbf{A} = \begin{bmatrix} a_{11} & a_{12}^T & a_{13} \\ a_{21} & A_{22} & a_{23} \\ a_{31} & a_{32}^T & a_{33} \end{bmatrix}, \tag{5.9.4}$$

in which A_{22} is $(n-2)$-by-$(n-2)$, and a_{11}, a_{33} are scalars. Define the matrices

$$\mathbf{B} = \begin{bmatrix} a_{11} & a_{12}^T \\ a_{21} & A_{22} \end{bmatrix}, \ \mathbf{C} = \begin{bmatrix} a_{12}^T & a_{13} \\ A_{22} & a_{23} \end{bmatrix}, \ \mathbf{D} = \begin{bmatrix} a_{21} & A_{22} \\ a_{31} & a_{32}^T \end{bmatrix}, \ \mathbf{E} = \begin{bmatrix} A_{22} & a_{23} \\ a_{32}^T & a_{33} \end{bmatrix}.$$

If we let $b = \det B$, $c = \det C$, $d = \det D$, and $e = \det E$, then it follows that

$$\det \begin{bmatrix} b & c \\ d & e \end{bmatrix} = \det A_{22} \det A.$$

Hence, provided $\det A_{22} \neq 0$, we have

$$\det A = \frac{\det B \det E - \det C \det D}{\det A_{22}}. \tag{5.9.5}$$

With certain nonnegativity/positivity assumptions **IM** matrices can be characterized in terms of Schur complements [JS01a].

Theorem 5.9.7 Let $A = [a_{ij}] \geq 0$. Then $A \in$ **IM** if and only if A has positive diagonal entries, all Schur complements are nonnegative, and all Schur complements of order 1 are positive.

In fact, these conditions can be somewhat relaxed.

Theorem 5.9.8 Let $A = [a_{ij}] \geq 0$. Then $A \in$ **IM** if and only if

(i) A has at least one positive diagonal entry,
(ii) all Schur complements of order 2 are nonnegative, and
(iii) all Schur complements of order 1 are positive.

Proof For necessity, assume that $A \in$ **IM**. Then A has positive diagonal entries and, because **IM** matrices are closed under extraction of Schur complements, each Schur complement is nonnegative. So we just need to show those of order 1 are positive. But this follows from Schur's formula because A has positive principal minors.

For sufficiency, suppose (i), (ii), and (iii) hold, say, $a_{ii} > 0$. Observe that (by considering all Schur complements $A[\{i,j\}]/a_{ii}$ in which $j \neq i$), (iii) implies that A has positive diagonal entries.

Claim. All principal minors of A are positive.

Proof of Claim. If A is 1-by-1, then the claim certainly holds. So, inductively, assume the claim holds for all matrices of order $<n$ satisfying (i), (ii), and (iii) and let $A = \begin{bmatrix} a_{11} & A_{12} \\ A_{21} & A_{22} \end{bmatrix}$. Thus, all principal submatrices of order $<n$ have positive determinant, and it suffices to prove that $\det A > 0$. By Schur's formula, $\det A = (\det A_{22})(a_{11} - A_{12}(A_{22})^{-1}A_{21})$. The inductive hypothesis implies

$\det A_{22} > 0$ and thus the positivity of $\det A$ follows from (iii), completing the proof of the claim.

Now let $A^{-1} = B = [b_{ij}]$ and consider b_{ij}, $i \neq j$, and assume, without loss of generality, that $i < j$. Define the sequences

$$\alpha = < 1, \ldots, i-1, i+1, \ldots, j-1, j+1, \ldots, n >,$$

$$\alpha_1 = < 1, \ldots, i-1, i+1, \ldots, j-1, j+1, \ldots, n, i >,$$

and

$$\alpha_2 = < 1, \ldots, i-1, i+1, \ldots, j-1, j+1, \ldots, n, j > .$$

Then,

$$\begin{aligned} b_{ij} &= (-1)^{i+j} \frac{\det A(j,i)}{\det A} \\ &= (-1)^{i+j}(-1)^{n-i-1}(-1)^{n-j} \frac{\det A[\alpha_1, \alpha_2]}{\det A} \\ &= -(a_{ij} - A[i,\alpha](A[\alpha])^{-1}A[\alpha, j]) \frac{\det A[\alpha]}{\det A} \\ &\leq 0. \end{aligned}$$

The latter inequality holds because $A/A[\alpha]$ is a Schur complement of order 2 and hence is nonnegative. Thus, $A^{-1} \in \mathbf{Z}$ which implies $A \in \mathbf{IM}$, and completes the proof. □

From the latter part of the proof of Theorem 5.9.8 we obtain another characterization of **IM** matrices.

Theorem 5.9.9 Let $A = [a_{ij}] \geq 0$. Then $A \in \mathbf{IM}$ if and only if

(i) $\det A > 0$ and
(ii) for each principal submatrix B of order $n-2$, $\det B > 0$ and $A/B \geq 0$.

As noted in Corollaries 5.9.2 and 5.9.3, **IM** matrices are closed under extraction of Schur complements and under extraction of principal submatrices. Conversely, if $A \geq 0$ with principal submatrix B and both B and A/B are **IM**, then A is not necessarily **IM**.

5.10 Submatrices

Summary. In this section almost principal submatrices are defined and signs of almost principal minors (*APM*s) of **M** and **IM** matrices are established. The operations of inverting and extracting a principal submatrix are shown to lead to entry-wise inequalities when these operations are applied to **M** and **IM**

matrices. For normalized **IM** matrices, it is shown that if a principal (resp., almost principal) submatrix is properly contained in another of the same type, then the magnitude of the larger principal (resp., almost principal) minor is smaller than that of the smaller, i.e., "larger" minors are smaller. It is also shown that if the inverse of a proper principal submatrix contains a block of zeros, then the inverse of the matrix itself has a larger block of zeros in related positions. Moreover, if a proper principle submatrix of an **IM** matrix contains a 0, then so does some principal submatrix of the same size. Finally, the last statement holds for general invertible matrices, and analogs of these results are shown to hold for positive definite matrices.

5.10.1 Principal and Almost Principal Submatrices

Square submatrices that are defined by index sets differing in only one index, or the minors that are their determinants, are called **almost principal submatrices**. For simplicity we abbreviate "almost principal minor" ("principal minor") to *APM* (*PM*). *APM*s are special for a variety of reasons including that, in the co-factor form of the inverse, they are exactly the numerators of off-diagonal entries of inverses of principal submatrices. So, if $A \in \mathbf{M}$ or $A \in \mathbf{IM}$, an *APM* is 0 if and only if an off-diagonal entry of the inverse of a principal submatrix equals 0. Using the informal notation $\alpha + i$ ($\alpha - i$) to denote the augmentation of the set α by $i \notin \alpha$ (deletion of $i \in \alpha$ from α), almost principal submatrices are of the form $A[\alpha + i, \alpha + j], i,j \notin \alpha$ ($A[\alpha - i, \alpha - j], i,j \in \alpha$), $i \neq j$. All *PM*s in $A \in \mathbf{M}$ or $A \in \mathbf{IM}$ are positive. Because of inheritance (of the property **IM** under extraction of principal submatrices [Joh82]), the sign of every nonzero *APM* in $A \in \mathbf{M}$ or $A \in \mathbf{IM}$ is determined entirely by its position. Specifically, if $\alpha \subseteq N$ and $i,j \in \alpha$, $sgn(\det A[\alpha - i, \alpha - j])$ equals $(-1)^{r+s}$ if $A \in \mathbf{M}$ and $(-1)^{r+s+1}$ if $A \in \mathbf{IM}$ in which r (resp., s) is the number of indices in α less than or equal to i (resp., j). An analogous statement can be made concerning $\det A[\alpha + i, \alpha + j]$. For an individual minor of an **IM** matrix that is neither principal nor an *APM*, there is no constraint upon the sign.

Our purpose here is to present more subtle information about *APM*s of an **IM** matrix: certain inequalities and relations among those that may be 0. To clarify our interest, consider the following example.

Example 5.10.1 Consider the **IM** matrix

$$\mathbf{A} = \begin{bmatrix} 1 & .5 & .4 & .2 \\ .8 & 1 & .8 & .4 \\ .6 & .5 & 1 & .4 \\ .2 & .2 & .25 & 1 \end{bmatrix}.$$

Notice that the only vanishing *APM*s are the determinants of $A[\{1,2\},\{2,3\}]$, $A[\{1,2\},\{2,4\}]$, $A[\{1,2,3\},\{2,3,4\}]$, and $A[\{1,2,4\},\{2,3,4\}]$. Thus, the $(1,3)$ entry of each of $A[\{1,2,3\}]^{-1}$ and $A[\{1,2,4\}]^{-1}$ is 0 while both the $(1,3)$ and $(1,4)$ entries of A^{-1} are 0 and these are the only entries that vanish in the inverse of any principal submatrix.

The preceding example leads to three questions.

Question 5.10.2 If the inverse of a proper principal submatrix of an **IM** matrix contains a block of 0s, does this imply that the inverse of the matrix itself has a larger block of 0s (and, somehow, in related positions)?

Question 5.10.3 If the inverse of an **IM** matrix contains a block of 0s, does this imply that there is a block of 0s in the inverse of some other principal submatrix?

Question 5.10.4 If the inverse of a proper principal submatrix of an **IM** matrix contains a 0, does this imply that the inverse of some other proper principal submatrix of the same size must also contain a 0?

The answers to Questions 5.10.2 and 5.10.4 are certainly not in the affirmative for invertible matrices in general.

Example 5.10.5 Consider the invertible matrix

$$\mathbf{A} = \begin{bmatrix} 1 & 2 & 4 & 3 \\ 3 & 1 & 2 & 1 \\ 2 & 3 & 4 & 1 \\ 4 & 3 & 2 & 5 \end{bmatrix}.$$

The $(3,1)$ minor of $A[\{1,2,3\}]$ is 0, but no other minor of A is 0. In fact, no other minor of any principal submatrix of A is 0.

Of course, there may be "isolated" 0s in the inverse of an **IM** matrix, and examples are easily found in which there is a single 0 entry in its inverse. All three questions above will be answered affirmatively later, with precise descriptions of these phenomena. Moreover, Question 5.10.3 will be answered for invertible matrices in general.

Because of Jacobi's determinantal identity, there are often analogous statements about matrices in **M**. It should be noted that besides *PM*s and *APM*s no other minors have deterministic signs throughout **M** or throughout **IM**; analogously, there seem to be no results like those we present beyond *PM*s and *APM*s.

5.10.2 Inverses and Principal Submatrices

The two operations of inverting and extracting a principal submatrix do not, of course, in general commute when applied to a given matrix (for which both are defined). There are, however, several interesting, elementary, entry-wise inequalities when these operations are applied to $A \in \mathbf{M}$ or $A \in \mathbf{IM}$ (for which they are always defined) in various orders. We record these here for reference and reflection. (Compare to inequalities in the positive semidefinite ordering for positive definite matrices [JS01b].) Throughout $\le$ or $<$ should be interpreted entry-wise.

Theorem 5.10.6 If $A \in \mathbf{M}$ or $A \in \mathbf{IM}$ and $\emptyset \neq \alpha \subseteq \langle n \rangle$, then

(i) $(A^{-1}[\alpha])^{-1} \le A[\alpha]$; and
(ii) $A[\alpha]^{-1} \le A^{-1}[\alpha]$.

Proof Notice that (i) holds for $A \in \mathbf{M}$ because

$$(A^{-1}[\alpha])^{-1} = A/A[\alpha^c] = A[\alpha] - A[\alpha,\alpha^c]A[\alpha^c]^{-1}A[\alpha^c,\alpha] \le A[\alpha].$$

(The latter inequality holds because $A[\alpha^c]^{-1} \ge 0$.) (i) holds for $A \in \mathbf{IM}$ by the same argument because $A[\alpha^c]^{-1}A[\alpha^c,\alpha] \ge 0$. (The former fact has been noted previously [JS01b], while the latter fact does not seem to have appeared in the literature.) Because (ii) for $A \in \mathbf{M}$ ($A \in \mathbf{IM}$) is just a restatement of (i) for $A \in \mathbf{IM}$ ($A \in \mathbf{M}$), the theorem holds. □

It follows from Theorem 5.10.6(ii) that if there are some 0 off-diagonal entries in $A[\alpha]^{-1}$, $A \in \mathbf{IM}$, then there are 0 entries in A^{-1} in the corresponding positions. In particular, if a certain *APM* in $A[\alpha]$ vanishes, then a larger (i.e., more rows and columns) *APM* in a corresponding position vanishes in A. Actually, more can be said, as we shall see later. A hint of this is the following. An entry of a square matrix is, itself, an *APM*; if an entry of $A \in \mathbf{IM}$ is 0, it is known that A must be reducible [JS11] and thus, if $n > 2$, other entries (i.e. other *APM*s of the same size) must be 0.

5.10.3 Principal and Almost Principal Minor Inequalities

It follows specifically from the observations of the last section that if $A \in \mathbf{IM}$, $\alpha \subseteq \beta \subseteq \langle n \rangle$, and the *APM* $\det A[\alpha+i,\alpha+j] = 0$, then $\det A[\beta+i,\beta+j] = 0$ when $i,j \notin \beta$. Thus, a "smaller" vanishing *APM* implies that any "larger" one containing it also vanishes. This suggests that there may, in general, be inequalities between such minors. We note, in advance, that Theorem 5.10.6(ii) gives some inequalities, but we give additional ones here.

We first give inequalities for the normalized case that may be paraphrased as saying that "larger" minors are smaller. The first inequality is a special case of Fischer's inequality, while the second was derived in [JS01b].

Theorem 5.10.7 If $A \in \mathbf{IM}$ is normalized and $\emptyset \neq \alpha \subseteq \beta \subseteq \langle n \rangle - \{i,j\}$, then

(i) $\det A[\beta] \leq \det A[\alpha]$; and
(ii) $|\det A[\beta + i, \beta + j]| \leq |\det A[\alpha + i, \alpha + j]|$.

Proof Noting that the determinantal inequalities of Fischer and Hadamard hold for inverse M-matrices [HJ91], we have $\det A[\beta] \leq \det A[\beta - \alpha]\ \det A[\alpha] \leq \det A[\alpha]$, which establishes (i).

Now assume that $i \neq j$, $B = A[\beta + i + j]$, and $\gamma = \alpha + i + j$. Then $B \in \mathbf{IM}$ also, and, from Theorem 5.10.6(ii), we have $|\left(B^{-1}\right)_{ij}| \leq |\left(B[\gamma]^{-1}\right)_{ij}|$ or equivalently,

$$\left|\frac{\det B[\beta + i, \beta + j]}{\det B}\right| \leq \left|\frac{\det B[\alpha + i, \alpha + j]}{\det B[\gamma]}\right|.$$

Thus,

$$\begin{aligned} |\det B[\beta + i, \beta + j]| &\leq \frac{\det B}{\det B[\gamma]} |\det B[\alpha + i, \alpha + j]| \\ &\leq (\det B[\beta + i + j - \gamma])\, |\det B[\alpha + i, \alpha + j]| \text{ (by Fischer)} \\ &\leq |\det B[\alpha + i, \alpha + j]| \text{ (by Hadamard).} \end{aligned}$$

Because B is a principal submatrix of A, (ii) follows. □

Because $A = [a_{ij}] \in \mathbf{IM}$ may be normalized via multiplication by $D = \operatorname{diag}(a_{11}, \ldots, a_{nn})^{-1}$, we may easily obtain nonnormalized inequalities from Theorem 4.10.4.

Theorem 5.10.8 If $A = [a_{ij}] \in \mathbf{IM}$ and $\emptyset \neq \alpha \subseteq \beta \subseteq \langle n \rangle - \{i,j\}$, then

(i) $\det A[\beta] \leq \det A[\alpha] \prod_{i \in \beta - \alpha} a_{ii}$; and
(ii) $|\det A[\beta + i, \beta + j]| \leq |\det A[\alpha + i, \alpha + j]| \prod_{i \in \beta - \alpha} a_{ii}$.

5.11 Vanishing Almost Principal Minors

From prior discussion we know that, for $A \in \mathbf{IM}$, if an entry of $A[\alpha]^{-1}$ is 0, then a corresponding entry of A^{-1} is 0. However, if $\alpha \subseteq \langle n \rangle$ properly, one quickly finds that, unlike for general matrices, it is problematic to construct an example in which an entry of $A[\alpha]^{-1}$ is 0 and just one entry of A^{-1} (the corresponding one) is 0. We present here two results that show there is a good reason for this. In one, it is shown that any block of 0s in $A[\alpha]^{-1}$ implies a "larger" block of 0s in A^{-1}, and in the other it is shown that any 0 entry in $A[\alpha]^{-1}$ implies 0s in certain other matrices $A[\beta]^{-1}$ when $|\beta| = |\alpha| < n$. Both results lead to interpretations in terms of rank of submatrices of **IM** matrices rather like the row/column inclusion results for positive semidefinite and other matrices, noted in [Joh98].

Because a 0 entry in $A \in$ **IM** implies that A is reducible and thus has other 0 entries (if $n > 2$), it follows that if $A \in$ **IM** and $A[\alpha]^{-1}$ has a 0 entry, $|\alpha| = 2$, then A^{-1} is reducible and the 0 entry is actually part of a 0 block in A^{-1}. This generalizes substantially even when $A \in$ **IM** is positive and answers Question 5.10.2.

Theorem 5.11.1 Suppose that $A \in$ **IM** and that $\gamma = \langle n \rangle - i$ for some $i \in \langle n \rangle$. Then, if $A[\gamma]^{-1}$ has a p-by-q 0 submatrix, A^{-1} has either a p-by-$(q+1)$ or a $(p+1)$-by-q 0 submatrix. Specifically, if $A[\gamma]^{-1}[\alpha, \beta] = 0$ in which $|\alpha| = p$ and $|\beta| = q$, then either $A^{-1}[\alpha, \beta + i] = 0$ or $A^{-1}[\alpha + i, \beta] = 0$.

Proof Let $A \in$ **IM** and $\gamma = \langle n \rangle - i$ for some $i \in \langle n \rangle$. Without loss of generality, assume that A has the partitioned form

$$\mathbf{A} = \begin{bmatrix} A_{11} & A_{12} \\ A_{21} & A_{22} \end{bmatrix} \tag{5.11.1}$$

in which $A_{11} = a_{ii}$ for some i, $1 \leq i \leq n$, and $A_{22} = A[\gamma]$. If an invertible matrix A is partitioned as in (5.11.1) with A_{22} invertible, then

$$\mathbf{A}^{-1} = \begin{bmatrix} s^{-1} & -s^{-1}u^T \\ -s^{-1}v & A_{22}^{-1} + s^{-1}vu^T \end{bmatrix} \tag{5.11.2}$$

in which $s = A/A_{22}$, $u^T = A_{12}A_{22}^{-1}$, and $v = A_{22}^{-1}A_{21}$. Thus, if $A_{22} \in$ **IM**, it is easily seen (and was first noticed in [Joh82]) that $A \in$ **IM** if and only if (i) $s > 0$, (ii) $u^T \geq 0$, (iii) $v \geq 0$, and (iv) $A_{22}^{-1} \leq -s^{-1}vu^T$, except for diagonal entries. Now $A_{22}^{-1} \in$ **M** and hence is in Z. Assume that A_{22}^{-1} has a p-by-q submatrix of 0s, say $A_{22}^{-1}[\alpha, \beta] = 0$ in which $|\alpha| = p$ and $|\beta| = q$. If $v_r = 0$ for all $r \in \alpha$, then $A^{-1}[\alpha, \beta + i]$ is a p-by-$(q+1)$ 0 submatrix of A^{-1}. On the other hand, if $v_r \neq 0$ for some $r \in \alpha$, then it follows from (iv) that $u_s = 0$ for all $s \in \beta$ and hence $A^{-1}[\alpha + i, \beta]$ is a p-by-$(q+1)$ 0 submatrix of A^{-1}. This completes the proof. □

In regard to Example 5.10.1, we see that for $\gamma = \{1, 2, 3\}$ or $\{1, 2, 4\}$, a 1-by-1 block of 0s in $A[\gamma]^{-1}$ leads to a 1-by-2 block of 0s in A^{-1}. It is easy to construct examples such that both possibilities for the 0 block of A^{-1} occur and also such that exactly one of the possibilities occurs.

Let the **measure of irreducibility**, $m(A)$, of an invertible n-by-n matrix A be defined as

$$m(A) = max_{(p,q)\in S}(p+q)$$

in which $S = \{ (p,q) \mid A \text{ contains a } p\text{-by-}q \text{ off-diagonal zero submatrix} \}$. Thus, an n-by-n matrix A is reducible if and only if $m(A) = n$. Moreover, in Theorem 5.11.1, we have shown that if B is an $(n-1)$-by-$(n-1)$ principal submatrix of $A \in$ **IM**, then $m\left(A^{-1}\right) \geq m\left(B^{-1}\right) + 1$. In general, if B is a k-by-k principal submatrix of $A \in$ **IM**, then $m\left(A^{-1}\right) \geq m\left(B^{-1}\right) + n - k$. (Note that the latter statement implies that if B is a reducible principal submatrix of $A \in$ **IM**, then A is reducible also. And, in particular, if $A \in$ **IM** has a 0 entry, then A is reducible, a fact noted previously.) The question is: when are these inequalities in fact equalities?

In order to establish a converse to Theorem 5.11.1 we will need the following result on **complementary nullities**. This fact has origins in [Gan59] and has been refined, for example, in [JL92]. Let Null(A) denote the **(right) null space** of a matrix A, nullity(A) denote the **nullity** of A, i.e., the dimension of Null(A), and rank(A) denote the **rank** of A.

Theorem 5.11.2 (Complementary Nullities) Let A be an n-by-n invertible matrix and $\emptyset \neq \alpha, \beta \subseteq \langle n \rangle$ with $\alpha \cap \beta = \emptyset$. Then, $\text{Null}(A^{-1}[\alpha,\beta]) = \text{Null}(A[\beta^c,\alpha^c])$.

We use Theorem 5.11.1 to prove a general result on 0 patterns of inverses.

Theorem 5.11.3 Let $\emptyset \neq \alpha, \beta \subseteq \langle n \rangle$ with $\alpha \cap \beta = \emptyset$, and let $A \in M_n(\mathbb{F})$ where $\mathbb{F}$ is an arbitrary field. If $A^{-1}[\alpha,\beta] = 0$, then, for any $\gamma \subseteq \langle n \rangle$ such that $\alpha \cap \gamma \neq \emptyset$, $\beta \cap \gamma \neq \emptyset$, $(\alpha \cup \beta)^c \subseteq \gamma$, and $A[\gamma]$ is invertible, we have $A[\gamma]^{-1}[\alpha \cap \gamma, \beta \cap \gamma] = 0$.

Proof Suppose that $A \in M_n(\mathbb{F})$ and $A^{-1}[\alpha,\beta] = 0$ in which $\emptyset \neq \alpha, \beta \subseteq \langle n \rangle$ with $\alpha \cap \beta = \emptyset$. Then $\text{nullity}(A^{-1}[\alpha,\beta]) = |\beta|$; hence, by complementary nullity, $\text{nullity}(A[\beta^c,\alpha^c]) = |\beta|$. Therefore,

$$\text{rank}(A[\beta^c,\alpha^c]) = |\alpha^c| - |\beta| = n - |\alpha| - |\beta| = |(\alpha \cup \beta)^c|.$$

Further, suppose that $\gamma \subseteq \langle n \rangle$ satisfies $\alpha \cap \gamma \neq \emptyset$, $\beta \cap \gamma \neq \emptyset$, $(\alpha \cup \beta)^c \subseteq \gamma$, and $A[\gamma]$ is invertible. Because $\gamma - \alpha \cap \gamma = \gamma \cap \alpha^c$ and

$$\gamma - \beta \cap \gamma = \gamma \cap \beta^c,$$

$$\text{rank}(A[\gamma][\gamma - \beta \cap \gamma, \gamma - \alpha \cap \gamma]) \leq \text{rank}\,(A[\beta^c,\alpha^c]) = |(\alpha \cup \beta)^c|.$$

Thus,

$$\begin{aligned}\text{nullity}(A[\gamma][\gamma-\beta\cap\gamma,\gamma-\alpha\cap\gamma]) &\geq |\gamma-\alpha\cap\gamma| - |(\alpha\cup\beta)^c| \\ &= |\beta\cap\gamma| + |(\alpha\cup\beta)^c| - |(\alpha\cup\beta)^c| = |\beta\cap\gamma|.\end{aligned}$$

Hence, by complementary nullity,

$$\text{nullity}(A[\gamma]^{-1}[\alpha\cap\gamma,\beta\cap\gamma]) \geq |\beta\cap\gamma|.$$

This last inequality must be an equality, i.e., $A[\gamma]^{-1}[\alpha\cap\gamma,\beta\cap\gamma]=0$, which completes the proof. □

We can make similar statements concerning the minors of A and $A[\gamma]$ even if A is singular.

It is easy to show that the condition $(\alpha\cup\beta)^c\subseteq\gamma$ is necessary. For instance, in Example 5.10.1, $A^{-1}[\alpha,\beta]=0$ in which $\alpha=\{1\}$ and $\beta=\{3,4\}$. But if $\gamma=\alpha\cup\beta$ (and thus $\alpha\cap\gamma\neq\emptyset$ and $\beta\cap\gamma\neq\emptyset$, while it is not the case that $(\alpha\cup\beta)^c\subseteq\gamma$), then $A[\gamma]^{-1}$ has no 0 entries.

We are now able to answer Question 5.10.3.

Corollary 5.11.4 Let $A\in\mathbf{IM}$ with $m\left(A^{-1}\right)=p+q$, say $A^{-1}[\alpha,\beta]=0$ in which $\emptyset\neq\alpha,\beta\subseteq\langle n\rangle$ with $|\alpha|=p$ and $|\beta|=q$. Further, let $\gamma=N-i$ for some $i\in N$, and assume that $\alpha\cap\gamma\neq\emptyset$ and $\beta\cap\gamma\neq\emptyset$. Then, $A[\gamma]^{-1}[\alpha\cap\gamma,\beta\cap\gamma]=0$ if and only if $(\alpha\cup\beta)^c\subseteq\gamma$.

Proof Assume the hypothesis holds and observe that $\alpha\cap\beta=\emptyset$ because $A\in\mathbf{IM}$. First, suppose that $A[\gamma]^{-1}[\alpha\cap\gamma,\beta\cap\gamma]=0$. Then $m\left(A^{-1}\right)\geq|\alpha\cap\gamma|+|\beta\cap\gamma|$. By the remarks after Theorem 5.11.1,

$$m\left(A^{-1}\right)\geq m\left(A[\gamma]^{-1}\right)+n-|\gamma|$$

and so

$$|\alpha|+|\beta|\geq|\alpha\cap\gamma|+|\beta\cap\gamma|+n-|\gamma|.$$

Rearranging, we have

$$(|\alpha|-|\alpha\cap\gamma|)+(|\beta|-|\beta\cap\gamma|)+|\gamma|\geq n$$

or equivalently,

$$|\alpha-\gamma|+|\beta-\gamma|+|\gamma|\geq n.$$

The latter holds (and with equality) if and only if $(\alpha\cup\beta)^c\subseteq\gamma$.

The converse follows from Theorem 5.11.3 because all principal submatrices of A are invertible. □

In regard to Example 5.10.1, $A^{-1}[\alpha,\beta] = 0$ in which $\alpha = \{1\}$ and $\beta = \{3,4\}$. For $\gamma = \{1,2,3\}$ or $\{1,2,4\}$, we have $\alpha \cap \gamma \neq \emptyset$, $\beta \cap \gamma \neq \emptyset$, and $(\alpha \cup \beta)^c \subseteq \gamma$. In each case $A[\gamma]^{-1}[\alpha \cap \gamma, \beta \cap \gamma] = 0$. On the other hand, for $\gamma = \{1,3,4\}$, we have $\alpha \cap \gamma \neq \emptyset$, $\beta \cap \gamma \neq \emptyset$, but $(\alpha \cup \beta)^c \subseteq \gamma$ does not hold, and $A[\gamma]^{-1}[\alpha \cap \gamma, \beta \cap \gamma] \neq 0$.

So we see that the inequality $m\left(A^{-1}\right) \geq m\left(B^{-1}\right)+1$ where B is an $(n-1)$-by-$(n-1)$ principal submatrix of A (noted in the discussion after Theorem 5.11.1) is an equality as long as $B = A[\gamma]$ in which $\alpha \cap \gamma \neq \emptyset$, $\beta \cap \gamma \neq \emptyset$, and $(\alpha \cup \beta)^c \subseteq \gamma$. Also, Example 5.10.1 (with $\gamma = \{2,3,4\}$) shows that, otherwise, the inequality may be strict.

We next discuss vanishing *APM*s and in response to Question 5.10.4 establish the fact that they imply that other *APM*s, of the same size, vanish. We will use the following lemma.

Lemma 5.11.5 Let $A = [a_{ij}]$ be an n-by-n **IM** matrix ($n \geq 3$) and $j \in \langle n \rangle$. If $a_{i_1j}, a_{i_2j}, \ldots, a_{i_tj} = 0$, then, for all $k \notin \{j, i_1, \ldots, i_t\}$, either $a_{kj} = 0$ or $a_{i_1k}, a_{i_2k}, \ldots, a_{i_tk} = 0$.

Proof This follows from the fact [Joh82, Wil77] that if A is an n-by-n **IM** matrix ($n \geq 3$), then for all $i,j,k \in \langle n \rangle$, $a_{ik}a_{kj} \leq a_{ij}a_{kk}$. □

Theorem 5.11.6 Let $A = [a_{ij}]$ be an n-by-n **IM** matrix ($n \geq 3$), let i,j,k be distinct indices in $\langle n \rangle$, and let δ be a subset of $\langle n \rangle - \{i,j,k\}$. Then, if $\det A[\delta + i, \delta + j] = 0$,

(i) $\det A[\delta + i, \delta + k] = 0$ or
(ii) $\det A[\delta + k, \delta + j] = 0$.

Proof Let $A = [a_{ij}]$ be an n-by-n **IM** matrix ($n \geq 3$), let i,j,k be distinct indices in $\langle n \rangle$, let δ be a subset of $\langle n \rangle - \{i,j,k\}$, and assume that $\det A[\delta + i, \delta + j] = 0$. If $\delta = \emptyset$, the result follows because $A[\{i,j,k\}]$ must be reducible. So assume that $\delta \neq \emptyset$. By permutation similarity we may assume that $i = i_1$, $\delta = \{i_2, \ldots, i_{p-1}\}$, and $j = i_p$. Let $A_1 = A[\{i_1, \ldots, i_p\}]$. Because $0 = \det A[\delta + i, \delta + j] = \det B$ where $B = A[\{i_1, \ldots, i_{p-1}\}, \{i_2, \ldots, i_p\}]$, we see that the (i_p, i_1) minor of A_1 is 0. Further, because $A[\delta]$ is a $(p-2)$-by-$(p-2)$ principal submatrix of A lying in the lower left corner of $B = [b_1, \ldots, b_{p-1}]$, $\{b_1, \ldots, b_{p-2}\}$ is linearly independent and thus $b_{p-1} = \sum_{i=1}^{p-2} \beta_i b_i$. If $b_{p-1} = 0$, then, by Lemma 5.11.5, either $a_{ki_p} = a_{kj} = 0$ or $a_{i_1k}, a_{i_2k}, \ldots, a_{i_{p-1}k} = 0$. The former case implies $\det A[\delta + k, \delta + j] = 0$, while the latter implies

$\det A[\delta + i, \delta + k] = 0$. So assume $b_{p-1} > 0$. Then $\beta_i \neq 0$ for some i, $1 \leq i \leq p-2$. By simultaneous permutation of rows and columns indexed by δ, we may assume $\beta_i \neq 0$.

Let $k = i_{p+1}$ and consider the submatrix $A_2 = A[\{i_1, \ldots, i_{p+1}\}]$ of A. Because A_1 is a principal submatrix of A_2, the (i_p, i_1) minor of A_2 is 0 also, i.e.,

$$0 = \det A[\{i_1, \ldots, i_{p-1}, i_{p+1}\}, \{i_2, \ldots, i_{p+1}\}] = \det A_3.$$

Now the $(p-1)$-by-$(p-1)$ submatrix lying in the upper left corner of A_3 is B and $\det B = 0$. Therefore, applying Sylvester's identity for determinants to $\det A_3$, we see that either

(1) $\det A[\{i_1, \ldots, i_{p-1}\}, \{i_3, \ldots, i_{p+1}\}] = 0$, or
(2) $\det A[\{i_2, \ldots, i_{p-1}, i_{p+1}\}, \{i_2, \ldots, i_p\}] = 0$.

Case I. Suppose (1) holds. Then,

$$\begin{aligned} 0 &= \det A[\{i_1, \ldots, i_{p-1}\}, \{i_3, \ldots, i_{p+1}\}] \\ &= \det[b_2, \ldots, b_{p-2}, b_{p-1}, b_p] \\ &= \det \left[b_2, \ldots, b_{p-2}, \sum_{i=1}^{p-2} \beta_i b_i, b_p \right] \\ &= \det[b_2, \ldots, b_{p-2}, \beta_1 b_1, b_p] \\ &= (-1)^{p-3} \beta_1 \det[b_1, b_2, \ldots, b_{p-2}, b_p]. \end{aligned}$$

Because $\beta_1 \neq 0$,

$$\begin{aligned} 0 &= \det[b_1, b_2, \ldots, b_{p-2}, b_p] \\ &= \det A[\{i_1, \ldots, i_{p-1}\}, \{i_2, \ldots, i_{p-1}, i_{p+1}\}] \\ &= \det A[\delta + i, \delta + k]. \end{aligned}$$

This establishes (i).

Case II. Suppose (2) holds. Then,

$$\begin{aligned} 0 &= \det A[\{i_2, \ldots, i_{p-1}, i_{p+1}\}, \{i_3, \ldots, i_{p+1}\}] \\ &= \det A[\delta + k, \delta + j]. \end{aligned}$$

This establishes (ii) and completes the proof. □

We can also prove Theorem 5.11.6 using Theorems 5.11.1 and 5.11.3 as follows. Let $\rho = \delta + i + j$ and $\tau = \rho + k$. Because $\det A[\delta + i, \delta + j] = 0$,

$A[\rho]^{-1}[i,j] = 0$. So, by Theorem 5.11.1, either $A[\tau]^{-1}[i,j+k] = 0$ or $A[\tau]^{-1}[i+k,j] = 0$. Suppose that $A[\tau]^{-1}[i,j+k] = A[\tau]^{-1}[\alpha,\beta] = 0$. Let $\gamma = \delta+i+k = \tau - j$. Then, in $A[\delta+i+j+k]$, we have $\alpha \cap \gamma \neq \emptyset$, $\beta \cap \gamma \neq \emptyset$, and $(\alpha \cup \beta)^c = \delta \subseteq \gamma$. Thus, by Theorem 5.11.3, $A[\tau]^{-1}[\alpha \cap \gamma, \beta \cap \gamma] = A[\gamma]^{-1}[i,k] = 0$, which implies that $\det A[\delta+i, \delta+k] = 0$. Similarly, it can be shown that if $A[\gamma]^{-1}[i+k,j] = 0$, then $\det A[\delta+k, \delta+j] = 0$.

In regard to Example 5.10.1, the $\{1,2\}$,$\{2,3\}$ minor of A vanishes. Hence, by Theorem 5.11.6, either the $\{1,2\}$,$\{2,4\}$ minor or the $\{2,4\}$,$\{2,3\}$ minor must vanish. The former was true.

M-matrices and **IM** matrices are known for their similarities with the positive definite matrices, and these results are no exception. In the positive definite matrices Theorem 5.11.6 has an obvious analog (namely that the $\delta+i, \delta+j$ minor is 0 if and only if the $\delta+j, \delta+i$ minor is 0), while the implication of Corollary 5.11.4 that follows from Theorem 5.11.3 has an identical analog. Less obvious is an analog to Theorem 5.11.1, yet there is one.

Recall that positive semidefinite matrices have an interesting property that may be called "row and column inclusion" [Joh98]. If A is an n-by-n positive semidefinite matrix, then, for any index set $\alpha \subseteq \langle n\rangle$ and any index $i \notin \langle n\rangle - \alpha$, the row $A[i,\alpha]$ (and thus the column $A[\alpha,i]$)lies in the row space of $A[\alpha]$ (column space of $A[\alpha]$). Of course, this is interesting only in the event that $A[\alpha]$ is singular, and, so, there is no complete analog for **IM** matrices in which every principal submatrix is necessarily invertible. However, just as there are slightly weakened analogs in the case of totally nonnegative matrices "row *or* column inclusion" [Joh98]), Theorem 5.11.1 implies certain analogs for **IM** matrices; now the role of principal submatrices is replaced by almost principal submatrices. We will need the following lemma. Here, Row(A) (Col(A)) denotes the row (column) space of a matrix A.

Lemma 5.11.7 Let the n-by-n matrix B have the partitioned form

$$\mathbf{B} = \begin{bmatrix} C & d \\ e^T & f \end{bmatrix}$$

in which f is a scalar and rank $(C) = n-2$. Then, if $d \notin Col(C)$ and $e^T \notin Row(C)$, B is invertible.

Proof Because $d \notin Col(C)$, rank $\left(\begin{bmatrix} C & d \end{bmatrix}\right) = n-1$ and because $e^T \notin Row(C)$, $\begin{bmatrix} e^T & f \end{bmatrix} \notin Row\left(\begin{bmatrix} C & d \end{bmatrix}\right)$. This implies rank$(B) = n$, i.e., B is invertible. □

Corollary 5.11.8 Let A be an n-by-n **IM** matrix, $\alpha \subseteq \langle n\rangle$, and $i,j \in \langle n\rangle - \alpha$. Then, for each $k \notin \alpha+i+j$, *either* $A[k,\alpha+j]$ lies in the row space of $A[\alpha+i,\alpha+j]$ *or* $A[\alpha+i,k]$ lies in the column space of $A[\alpha+i,\alpha+j]$.

Proof The corollary certainly holds if $A[\alpha + i, \alpha + j]$ is invertible. So assume not. Then $\det A[\alpha + i, \alpha + j] = 0$ and rank $(A[\alpha + i, \alpha + j]) = |\alpha|$. Thus, $A[\alpha + i + j]^{-1}$ has a 0 in the (i,j) position, and if $k \notin \alpha + i + j$, it follows from Theorem 5.11.1 that $A[\alpha + i + j + k]^{-1}$ also has a 0 in the (i,j) position. Hence, $\det A[\alpha + i + k, \alpha + j + k] = 0$. With $B = A[\alpha + i + k, \alpha + j + k]$ and $C = A[\alpha + i, \alpha + j]$, we see that the result follows upon applying Lemma 5.11.7. □

We note that in the statement of Corollary 5.11.8 the almost principal submatrix $A[\alpha + i, \alpha + j]$ could be replaced by $A[\alpha, \alpha + j]$ in the row case and by $A[\alpha + i, \alpha]$ in the column case to yield a stronger, but less symmetric, statement.

In the row or column inclusion results for totally nonnegative matrices, a bit more is true, the either/or statement is validated either always by rows or always by columns (or both), and this is (trivially) also so in the positive semidefinite case. Thus, it is worth noting that such phenomenon does *not* carry over to the almost principal **IM** case.

Example 5.11.9 Consider the **IM** matrix

$$\mathbf{A} = \begin{bmatrix} 14 & 4 & 1 & 1 & 4 \\ 6 & 16 & 4 & 4 & 6 \\ 6 & 6 & 14 & 4 & 6 \\ 6 & 6 & 4 & 14 & 6 \\ 4 & 4 & 1 & 1 & 14 \end{bmatrix}.$$

The hypothesis of the corollary is satisfied with $\alpha = \{2\}$, $i = 1$, and $j = 3$, and the conclusion is satisfied for $k = 4$ by columns and not rows and for $k = 5$ by rows and not columns.

Examples in which satisfaction is always via columns or always via rows are easily constructed.

5.12 The Path Product Property

Summary. In this section, for nonnegative matrices, the path product (**PP**) property is defined as well as the strict path product (**SPP**) property. It is shown that every **IM** matrix is **SPP** and, for $n \leq 3$, every **SPP** matrix is **IM**. However, the latter statement does not hold for $n > 3$. It is noted that every normalized **PP** (normalized **SPP**) matrix A has a normalized matrix $\hat{A}$ that is positive diagonally similar to A in which all (resp., off-diagonal) entries are ≤ 1 (resp. < 1). A characterization of invertible, triangular, normalized **SPP** matrices is given in terms of path products, thereby answering a question posed in [Joh82]. It is noted that **PP** matrices can be used to establish a number of facts about **IM**

matrices. The notions of purely strict path product (**PSPP**) and totally strict path product (**TSPP**) are introduced.

5.12.1 (Normalized) PP and SPP Matrices

Let $A = [a_{ij}]$ be an n-by-n nonnegative matrix with positive diagonal entries. We call A a **path product** (**PP**) matrix if, for any triple of indices $i, j, k \in N$,

$$\frac{a_{ij}a_{jk}}{a_{jj}} \leq a_{ik}$$

and a **strict path product** (**SPP**) matrix if there is strict inequality whenever $i \neq j$ and $k = i$ [JS01a]. In [JS01a] it was noted that any **IM** matrix is **SPP** and that for $n \leq 3$ (but not greater) the two classes are the same. See also [Wil77].

For a **PP** matrix A and any path $i_1 \to i_2 \to \cdots \to i_{k-1} \to i_k$ in the complete graph K_n on n vertices, we have

$$\frac{a_{i_1 i_2}a_{i_2 i_3} \cdots a_{i_{k-1} i_k}}{a_{i_2 i_2}a_{i_3 i_3} \cdots a_{i_{k-1} i_{k-1}}} \leq a_{i_1 i_k} \tag{5.12.1}$$

and, if in addition, A is **SPP**, then the inequality is strict. We call (5.12.1) the **path product inequalities** and, if $i_1 = i_k$ in (5.12.1), the **cycle product inequalities**. We call a product $a_{i_1 i_2}a_{i_2 i_3} \cdots a_{i_{k-1} i_k}$ an (i_1, i_k) **path product** (of length $k-1$) and, if $i_k = i_1$, an (i_1, i_k) **cycle product** (of length $k-1$).

In (5.12.1), if $i_k = i_1$, we see that the product of entries around any cycle is no more than the corresponding diagonal product, i.e.,

$$a_{i_1 i_2}a_{i_2 i_3} \cdots a_{i_{k-1} i_1} \leq a_{i_1 i_1}a_{i_2 i_2} \cdots a_{i_{k-1} i_{k-1}}. \tag{5.12.2}$$

It follows that, in a normalized **PP** matrix, no cycle product is more than 1 and that, in a strictly normalized **PP** matrix, all cycles of length two or more have product less than 1. From this we have [JS01a].

Theorem 5.12.1 If A is a normalized **PP** (resp. normalized **SPP**) matrix, then there is a normalized **PP** (resp. normalized **SPP**) matrix $\hat{A}$ positive diagonally similar to A in which all (resp. off-diagonal) entries are $\leq$ (resp. $<$) 1.

PP matrices are closed under: extraction of principal submatrices, permutation similarity, Hadamard (entry-wise) multiplication, left (right) multiplication by a positive diagonal matrix (and hence positive diagonal congruence), and positive diagonal similarity but not under Schur complementation, addition, or ordinary multiplication [JS01a]. Moreover, a **PP** matrix remains **PP** upon the addition of a nonnegative diagonal matrix.

There is a strong connection between **IM** matrices and **SPP** matrices as the next several results from [JS01a] illustrate.

Theorem 5.12.2 Every **IM** matrix is **SPP**.

Theorem 5.12.3 If A is an n-by-n **SPP** matrix, $n \leq 3$, then A is **IM**.

For $n \geq 4$, sufficiency no longer holds.

Example 5.12.4 The **SPP** matrix

$$\mathbf{A} = \begin{bmatrix} 1 & 0.10 & 0.40 & 0.30 \\ 0.40 & 1 & 0.40 & 0.65 \\ 0.10 & 0.20 & 1 & 0.60 \\ 0.15 & 0.30 & 0.60 & 1 \end{bmatrix}$$

is not **IM**, as the $(2,3)$ entry of A^{-1} is positive.

Given an n-by-n matrix A, $G(A)$, the **directed graph of A**, is the graph with vertices $\langle n \rangle$ and satisfying: (i,j) is an edge of $G(A)$ if and only if $a_{ij} \neq 0$. A directed graph G with vertex set $V(G)$ and edge set $E(G)$ is a **transitive directed graph** if $(i,j), (j,k) \in E(G)$ implies $(i,k) \in E(G)$. For an n-by-n matrix A, we say A is a **transitive matrix** if $G(A)$ is transitive. **PP** matrices can be used to deduce the following important facts about **IM** matrices [JS01a], some of which we will use later.

Theorem 5.12.5

(i) If an **IM** matrix has a 0 entry, then it is reducible.
(ii) An **IM** matrix has a transitive graph.
(iii) An **IM** matrix can be scaled by positive diagonal matrices D, E so that DAE has all diagonal entries equal to 1 and all off-diagonal entries less than 1.
(iv) Every **IM** matrix satisfies the strict cycle product inequalities

$$a_{i_1 i_2} a_{i_2 i_3} \dots a_{i_{k-1} i_1} < a_{i_1 i_1} a_{i_2 i_2} \dots a_{i_{k-1} i_{k-1}}. \tag{5.12.3}$$

In fact, it follows from Theorem 5.12.5(i) that every 0 entry of an **IM** matrix lies in an off-diagonal reducing block. The known fact (see [JS01a, section 4.6], and references therein) that the 0-pattern of an **IM** matrix is power-invariant also follows from this discussion. Each is simply part of a reducing 0-block.

In [JS01a] it was shown that **SPP** matrices are not necessarily P-matrices in contrast to M-matrices and inverse M-matrices.

A path product $a_{i_1 i_2} a_{i_2 i_3} \dots a_{i_{k-1} i_k}$ is called an (i_1, i_k) **path product**, and we say it is an *even* (*odd*) if the path $i_1 \to i_2 \to \cdots \to i_{k-1} \to i_k$ in $G(A)$ has even (odd) length (equivalently, if k is odd(even)). Note that if $A \geq 0$, then the (i,j) entry of A^k, $k = 1, 2, \ldots,$ cquals the sum of the (i,j) path products of length k.

Theorem 5.12.6 Let A be an n-by-n invertible, triangular, normalized **SPP** matrix. Then, $A \in$ **IM** if and only if the sum of the even length (i,j) path products is at most the sum of the odd length (i,j) path products for all i and j with $i \neq j$.

Proof Let $A = I + T$ be an n-by-n invertible, triangular, normalized **SPP** matrix. Then,

$$\begin{aligned} A^{-1} &= (I+T)^{-1} \\ &= I - T + T^2 - \cdots \pm T^{n-1} \\ &= I - \sum_{k\, odd} T^k + \sum_{k\, even} T^k. \end{aligned}$$

Thus, we see that, for all i and j with $i \neq j$, the (i,j) entry of A^{-1} is the sum of the even (i,j) path products minus the sum of the odd (i,j) path products. Hence, $A^{-1} \in$ **M** if and only if the sum of the even (i,j) path products is at most the sum of the odd (i,j) path products for all i and j with $i \neq j$. □

5.12.2 TSPP and PSPP Matrices

We identify the case in which no path product equalities occur, i.e., all inequalities (5.12.1) are strict, and call such **SPP** matrices **totally strict path product matrices** (**TSPP**) [JS07b]. Observe that **TSPP** matrices are necessarily positive, but may not be **IM** and that **IM** matrices may not be **TSPP**.

Example 5.12.7 The **TSPP** matrix

$$\mathbf{A} = \begin{bmatrix} 1 & 0.50 & 0.35 & 0.40 \\ 0.50 & 1 & 0.50 & 0.26 \\ 0.35 & 0.50 & 1 & 0.50 \\ 0.40 & 0.26 & 0.50 & 1 \end{bmatrix}$$

is not **IM** because the $(2,4)$ entry of the inverse is positive while the **IM** matrix

$$\mathbf{B} = \begin{bmatrix} 1 & 0.50 & 0.25 \\ 0.50 & 1 & 0.50 \\ 0.25 & 0.50 & 1 \end{bmatrix}$$

is not **TSPP**.

Consider the following condition on the collection of path product inequalities: for all distinct indices $i, j, k \in \langle n \rangle$ and, for all $m \in \langle n \rangle - \{i, j, k\}$,

$$a_{ik} = a_{ij}a_{jk} \quad \text{implies that either} \quad a_{im} = a_{ij}a_{jm} \quad \text{or} \quad a_{mk} = a_{mj}a_{jk}. \quad (5.12.4)$$

If (5.12.4) is satisfied by an **SPP** matrix, we say that A is a **purely strict path product matrix** (**PSPP**) [JS01a, JS07b]. We will see (Section 5.16) that, in **PSPP** matrices, path product equalities force certain cofactors to vanish.

We note that **PSPP** and **SPP** coincide (vacuously) when $n \leq 3$ and that generally the **TSPP** matrices are contained in the **PSPP** matrices (vacuously). Also observe that, if A is **TSPP** (**PSPP**), then so is any normalization of A. Last, we note that an **IM** matrix is necessarily **PSPP** (this was proved for positive normalized **IM** matrices in [JS01a], and the same proof applies in the general case), while a **TSPP** matrix is not necessarily **IM**; see Example 5.12.7, for instance.

5.13 Triangular Factorizations

In considering triangular factorizations, first note that **PP** matrices do not necessarily admit an LU-factorization in which the factors are **PP**.

Example 5.13.1 The **PP** matrix

$$\mathbf{A} = \begin{bmatrix} 1 & 0 & 0 & 1 \\ 1 & 1 & 1 & 1 \\ 0 & 0 & 1 & 1 \\ 0 & 0 & 0 & 1 \end{bmatrix}$$

has LU factorization

$$\mathbf{A} = \begin{bmatrix} 1 & 0 & 0 & 0 \\ 1 & 1 & 0 & 0 \\ 0 & 0 & 1 & 0 \\ 0 & 0 & 0 & 1 \end{bmatrix} \begin{bmatrix} 1 & 0 & 0 & 1 \\ 0 & 1 & 1 & 0 \\ 0 & 0 & 1 & 1 \\ 0 & 0 & 0 & 1 \end{bmatrix},$$

but $U = [u_{ij}]$ is not **PP** because $u_{23}u_{34} \not\leq u_{24}$.

However, M-matrices and inverse M-matrices have LU- (UL-) factorizations within their respective classes. To see this, first suppose that $A \in \mathbf{M}$. If A is 1-by-1, A trivially factors as LU in which $L, U > 0$. So assume $n > 1$. Because M-matrices are closed under extraction of principal submatrices, we may assume (inductively) that $A = \begin{bmatrix} A_{11} & A_{12} \\ A_{21} & A_{22} \end{bmatrix}$ in which $A_{11} = L_{11}U_{11}$ with L_{11} (U_{11})

being an $(n-1)$-by-$(n-1)$ lower- (upper-) triangular M-matrix. Therefore, because A has positive principal minors

$$A = LU = \begin{bmatrix} L_{11} & 0 \\ L_{21} & 1 \end{bmatrix} \begin{bmatrix} U_{11} & U_{12} \\ 0 & u_{nn} \end{bmatrix} = \begin{bmatrix} L_{11}U_{11} & L_{11}U_{12} \\ L_{21}U_{11} & L_{21}U_{12} + u_{nn} \end{bmatrix}$$

in which $u_{nn} = A/A_{11} > 0$ (by Schur's formula). Thus, $L_{11}U_{12}, L_{21}U_{11} \leq 0$. Because $L_{11}^{-1}, U_{11}^{-1} \geq 0$, it follows that $U_{12}, L_{21} \leq 0$. Thus, $L, U \in Z$. And because

$$L^{-1} = \begin{bmatrix} L_{11}^{-1} & 0 \\ -L_{21}L_{11}^{-1} & 1 \end{bmatrix} \quad \text{and} \quad U^{-1} = \begin{bmatrix} U_{11}^{-1} & -\frac{1}{u_{nn}}U_{11}^{-1}U_{12} \\ 0 & \frac{1}{u_{nn}} \end{bmatrix}$$

are both nonnegative, $L, U \in \mathbf{M}$. Similarly, it can be shown that A has a UL-factorization within the class of M-matrices. Observe that if $A = LU$ (UL) in which $L, U \in \mathbf{M}$, then $A^{-1} = U^{-1}L^{-1}$ $(L^{-1}U^{-1})$ with $L^{-1}, U^{-1} \in \mathbf{IM}$. Thus, **IM** matrices also have LU- and UL-factorizations within their class. However, it is not the case that if L, U are, respectively, lower- and upper-triangular **IM** matrices, then LU and/or UL is **IM**.

Example 5.13.2 Consider the **IM** matrices

$$L = \begin{bmatrix} 1 & 0 & 0 \\ 1 & 1 & 0 \\ 4 & 1 & 1 \end{bmatrix}, \quad U = \begin{bmatrix} 1 & 1 & 4 \\ 0 & 1 & 4 \\ 0 & 0 & 1 \end{bmatrix}.$$

Neither LU nor UL is **IM** because the inverse of the former is positive in the $(2, 1)$ entry, and the inverse of the latter is positive in the $(3, 2)$ entry.

5.14 Sums, Products, and Closure

Summary. In this section it is shown that sums, products, and positive integer powers of **IM** matrices need not be **IM** and are, in fact, **IM** if and only if the off-diagonal entries of its inverse are nonpositive. It is also noted that if $A \in \mathbf{IM}$, then $A^t \in \mathbf{IM}$ for all $0 \leq t < 1$. Throughout, let $A_1, A_2, \ldots, A_k$ be n-by-n **IM** matrices in which k is a positive integer.

It is natural to ask about closure under a variety of operations. Neither the product AB nor the sum $A + B$ need generally be in **IM**; see Section 5.2. In addition, the conventional powers A^t, $t > 1$, are nonnegative but need not be in **IM** for $n > 3$. Recall that A^t, when t is not an integer, is defined naturally for an M-matrix, via power series, and thus for an **IM** matrix, as in Section 5.8, in [Joh82], or, equivalently, via principal powers, as in [JS11].

Example 5.14.1 Consider the **IM** matrix

$$\mathbf{A} = \begin{bmatrix} 90 & 59 & 44 & 71 \\ 56 & 96 & 48 & 88 \\ 64 & 60 & 88 & 84 \\ 42 & 43 & 36 & 95 \end{bmatrix}.$$

Because the inverse of A^3 has a positive 1, 4 entry, A^3 is not an **IM** matrix.

However, in each of these cases, there is an aesthetic condition for the result to be **IM**, even when the number k of summands or factors is more than two. Note that, because $A_1, A_2, \ldots, A_k \geq 0$ (entry-wise), necessarily $A_1 + A_2 + \cdots + A_k, A_1A_2 \ldots A_k \geq 0$. The latter of these is necessarily invertible, while the first need not be. But, invertibility plus nonnegativity means that the result is **IM** if and only if the inverse has nonpositive off-diagonal entries. This gives

Theorem 5.14.2 Let $A_1, A_2, \ldots, A_k$ be n-by-n **IM** matrices and let k be a positive integer. Then, $A_1 + A_+ \cdots + A_k$ (resp. $A_1A_2 \ldots A_k$), provided it is invertible, is **IM** if and only if the off-diagonal entries of its inverse are nonpositive.

Also,

Theorem 5.14.3 If $t \geq 1$ is an integer, then A^t is **IM** if and only if the off-diagonal entries of the inverse of A^t are nonpositive.

The remaining issue of "conventional" powers (roots) A^t, $0 < t < 1$, has an interesting resolution. Based upon a power series argument given in [Joh82] and in Section 5.8, the M-matrix $B = A^{-1}$ always has a k-th root $B^{1/k} \in$ **IM** (that is, $(B^{1/k})^k = B$), $k = 1, 2, \ldots$, that is an M-matrix.

The same argument or continuity gives a natural $B^T = \left(A^{-1}\right)^T$, $0 \leq t < 1$, that is an M-matrix. As the laws of exponents are valid, this means

Theorem 5.14.4 If A is **IM**, then for each t, $0 \leq t < 1$, there is a natural A^t such that A^t is **IM**. If $t = p/q$, with p and q positive integers and $0 \leq p \leq q$, then

$$\left(A^t\right)^q = A^p.$$

5.15 Spectral Structure of IM Matrices

Let $A \in$ **IM**, and let $\sigma(A)$ denote the spectrum of A. It follows from Perron–Frobenius theory that $|\lambda| \leq \rho(A)$ for all $\lambda \in \sigma(A)$ with equality only for

$\lambda = \rho(A)$ and that the M-matrix $B = A^{-1} = \alpha I - P$ in which $P \geq 0$ and $\alpha > \rho(P)$. Thus,

$$q(B) \equiv \frac{1}{\rho(A)} = \alpha - \rho(P)$$

is the eigenvalue of B with minimum modulus, $\sigma(B)$ is contained in the disc $\{z \in \mathbb{C}\colon |z - \alpha| \leq \rho(P)\}$, and $Re(\lambda) \geq q(B)$ for all $\lambda \in \sigma(B)$ with equality only for $\lambda = q(B)$. Moreover, $\sigma(B)$ is contained in the open wedge

$$W_n \equiv \left\{z = re^{i\theta} : r > 0, |\theta| < \frac{\pi}{2} - \frac{\pi}{n}\right\}$$

in the right half-plane if $n > 2$, and in $(0, \infty)$ if $n = 2$ [HJ91]. So $\sigma(A)$ is contained in the wedge W_n if $n > 2$ and in $(0, \infty)$ if $n = 2$. Under the transformation $f(z) = \frac{1}{z}$, circles are mapped to circles and lines to lines (see [MH06] for details). This in turn implies that $\sigma(A)$ is contained in the disc $\{z \in C\colon |z - \beta| \leq R\}$ in which $\beta = \frac{\alpha}{\alpha^2 - (\rho(P))^2}$ and $R = \frac{\rho(P)}{\alpha^2 - (\rho(P))^2}$.

5.16 Hadamard Products and Powers, Eventually IM Matrices

Summary. For m-by-n matrices $A = [a_{ij}]$ and $B = [b_{ij}]$, the **Hadamard (entry-wise) product** $A \circ B$ is defined by $A \circ B = [a_{ij}b_{ij}]$. Many facts about the Hadamard product may be found in [Hor90, Joh82].

In this section, it is shown that **IM** matrices are closed under Hadamard product if and only if $n \leq 3$. Also, it is shown that, for an **IM** matrix A, the t-th **Hadamard power** $A^{(t)} = [a_{ij}^t]$ is **IM** for all real $t \geq 1$ [Che07a], and that this is not necessarily the case if $n > 3$ and $0 < t < 1$. The dual of the class of **IM** matrices is discussed. In [JS01a], it was noted that any **IM** matrix is **SPP** and that for $n \leq 3$ (but not greater) the two classes are the same. See also [Wil77]. Our interest here lies in further relating these two classes via consideration of Hadamard powers. Motivated, in part, by the main result of [Che07a], we questioned whether an **SPP** matrix A becomes **IM** as t increases without bound. Toward this end, **IM** matrices are shown to be *PSPP* (see Section 5.12). Further, it is shown that **TSPP** matrices (those **SPP** matrices in which *all* path products are strict) become inverse M-matrices as t increases without bound but not conversely. However, the **PSPP** condition is shown to be necessary and sufficient to ensure that $A^{(t)}$ eventually becomes **IM**. In the process the notion of **path product triple** is introduced for an **IM** matrix A. (The presence of such a triple implies that a certain entry of the inverse of A vanishes.) We begin by establishing the fact that **IM** matrices are closed under Hadamard product for $n \leq 3$.

5.16.1 Hadamard Products and Powers

Because n-by-n **IM** matrices A and B are entry-wise nonnegative, it is natural to ask whether $A \circ B$ is again **IM**. For $n \leq 3$, this is so, as **IM** is equivalent to **SPP** for $n = 3$ (Section 5.12), and the Hadamard product of **SPP** matrices is **SPP**; and for $n = 1, 2$, the claim is immediate. It has long been known that such a Hadamard product is not always **IM**. Examples for $n = 6$ (and hence for larger n) may be [Joh77, Joh78], and more recently, it was noted in [WZZ00] that the two 4-by-4 symmetric matrices

$$A = \begin{bmatrix} 1 & 1 & 1 & 1 \\ 1 & 2 & 2 & 2 \\ 1 & 2 & 3 & 3 \\ 1 & 2 & 3 & 4 \end{bmatrix} \text{ and } B = \begin{bmatrix} 3 & 2 & 1 & 3 \\ 2 & 2 & 1 & 2 \\ 1 & 1 & 1 & 1 \\ 3 & 2 & 1 & 4 \end{bmatrix}$$

are **IM**, while $A \circ B$ is not. Thus, the Hadamard product, $A \circ B$, of two **IM** matrices A and B need not be **IM** for $n > 3$, which entirely resolves the question of dimensions in which there is Hadamard product closure.

Theorem 5.16.1 The n-by-n **IM** matrices are closed under Hadamard product if and only if $n \leq 3$.

This leaves the question of which pairs of **IM** matrices have an **IM** Hadamard product. Counterexamples seem not so common, in part because the Hadamard product is **SPP**, but also because the ideas in [JS07a] indicate that a bounded multiple of I (at worst) need be added to make the Hadamard product **IM**. Nonetheless, better descriptions of such pairs would be of interest. In [CFJ01, JS07a] the **dual of the IM matrices** was defined to be

$$\mathbf{IM}^D = \{A \in M_n(R) \colon A \circ B \in \mathbf{IM} \text{ for all } B \in \mathbf{IM}\}.$$

It is not hard to see that $\mathbf{IM}^D \subseteq \overline{\mathbf{IM}}$, but an effective characterization of $\mathbf{IM}^D$ would also be of interest.

It was noted in [Joh77, FP62] that if A and B are M-matrices, then the Hadamard product $A \circ B^{-1}$ is also an M-matrix. A real n-by-n matrix A is **diagonally symmetrizable** if there exists a diagonal matrix D with positive diagonal entries such that DA is symmetric. In [Joh77] it was shown that if the M-matrix A is diagonally symmetrizable, then $q\left(A \circ A^{-1}\right) = 1$. For an M-matrix A, there has been a great deal of interest in bounds for $q\left(A \circ A^{-1}\right)$. For instance, $q\left(A \circ A^{-1}\right) \leq 1$ was proved in [FJMN85], and moreover, it was asked whether $q\left(A \circ A^{-1}\right) \geq \frac{1}{n}$. This latter question was answered affirmatively in [FJMN85], and the authors conjectured that $q\left(A \circ A^{-1}\right) \geq \frac{2}{n}$. Also, for M-matrices A and B, a lower bound for $q\left(A \circ B^{-1}\right)$ was determined. The conjecture in [FJMN85] was later established independently in [Son00, Yon00, Che04b].

For $A = [a_{ij}] \in \mathbf{IM}$, another (more special) natural question is whether $A^{(2)} \equiv A \circ A \in \mathbf{IM}$ also [Neu98]. This was also conjectured elsewhere. A constructive proof for $A^{(2)}$ when $n = 4$ was given by the authors. More generally, is $A^{(k)} = A \circ A \circ \cdots \circ A \in \mathbf{IM}$ for all positive integers k? This question was settled in [Che04a].

Theorem 5.16.2 All positive integer Hadamard powers of an **IM** matrix are **IM**.

In [WZZ00], as well as elsewhere, it was then conjectured that, for an **IM** matrix A, the ***t*-th Hadamard power** of A is **IM** for all real $t \geq 1$. This was then proven in [Che07a].

Theorem 5.16.3 If A is an **IM** matrix and $t \geq 1$, $A^{(t)}$ is **IM**.

For $n \geq 4$ and $0 < t < 1$, Hadamard powers $A^{(t)}$ need not be **IM**.

Example 5.16.4 Consider the **IM** matrix

$$\mathbf{A} = \begin{bmatrix} 1 & 0.001 & 0.064 & 0.027 \\ 0.064 & 1 & 0.064 & 0.274625 \\ 0.001 & 0.008 & 1 & 0.216 \\ 0.003375 & 0.027 & 0.216 & 1 \end{bmatrix}.$$

The Hadamard cube root of A, given by

$$\mathbf{A}^{(1/3)} = \begin{bmatrix} 1 & 0.1 & 0.4 & 0.3 \\ 0.4 & 1 & 0.4 & 0.65 \\ 0.1 & 0.2 & 1 & 0.6 \\ 0.15 & 0.3 & 0.6 & 1 \end{bmatrix},$$

is not **IM** because the 2, 3 entry of its inverse is positive.

The question of which **IM** matrices satisfy $A^{(t)}$ is **IM** for all $t > 0$ is still open.

5.16.2 Eventually IM Matrices

As noted in Section 5.12, if A is **SPP** and $t > 0$, it is clear that $A^{(t)}$ is **SPP**. In fact, a Hadamard product of any two **SPP** matrices is again **SPP**, and this remains so for the variants of **SPP** to be studied in this section. This raises the question as to whether $A^{(t)}$ eventually becomes **IM** as t increases without bound. That is, is there a $T \geq 0$ such that $A^{(t)}$ is **IM** for all $t > T$? In this case we say that A is an **eventually inverse M-matrix** (**EIM**). Our question, then, is: which nonnegative matrices are **EIM**? It is clear that it is necessary that an **EIM** matrix be **SPP**, but it is not sufficient.

Example 5.16.5 Consider the normalized **SPP** matrix

$$\mathbf{A} = \begin{bmatrix} 1 & 0.5 & 0.7 & 0.4 \\ 0.5 & 1 & 0.5 & 0.25 \\ 0.7 & 0.5 & 1 & 0.5 \\ 0.4 & 0.25 & 0.5 & 1 \end{bmatrix}.$$

Because the 2,4 cofactor of $A^{(t)}$ is $c_{24}^{(t)} = [(0.5)^t - (0.35)^t][(0.4)^t - (0.35)^t]$, which is positive for all $t > 0$, we see that A is not **EIM**.

We also address the question of whether there is some converse to the statement that **IM** implies (some kind of) **SPP**.

In order to answer these questions, we refine further the path product conditions for **IM** matrices and identify the appropriate additional necessary conditions. These involve the arrangement of occurrences of equality in the path product inequalities (5.12.4). We say that (i,j,k) is a **path product equality triple** (for the **SPP** matrix A) if equality occurs in (5.12.4). For instance, $(2,3,4)$ is a path product equality triple in Example 5.16.6. Note that the path product equality triples of A and a normalized version of A are the same. When $n \geq 4$, we will see that path product equalities in an **IM** matrix (which can occur) necessarily imply others. In view of previous work [JS01b], this is not surprising, as a path product equality implies that a 2-by-2 almost principal minor is 0, so that a certain 3-by-3 principal submatrix has a 0 in its inverse. By [JS01b] this implies further 0s in the inverse of the full matrix (though it is not necessary that all inverse 0s stem from path product equalities).

Example 5.16.6 Consider the **IM** matrix

$$\mathbf{A} = \begin{bmatrix} 1 & 0.4 & 0.4 & 0.3 \\ 0.4 & 1 & 0.5 & 0.5 \\ 0.4 & 0.4 & 1 & 0.6 \\ 0.4 & 0.4 & 0.4 & 1 \end{bmatrix}.$$

Matrix A contains no path product equalities. However, A^{-1} has a 0 in the 1,4 position.

It was noted in [JS01a, theorem 4.3.1] that if A is an n-by-n **SPP** matrix, then there exist positive diagonal matrices D and E such that $B = DAE$ in which B is a normalized **SPP** matrix. Thus, $B^{(t)} = D^tA^{(t)}E^t$ for $t \geq 0$ and, because it is apparent that **SPP** matrices are closed under positive diagonal equivalence,

$A^{(t)}$ is **SPP** if and only if $B^{(t)}$ is. Therefore, it suffices to consider normalized **SPP** matrices when studying (positive) Hadamard powers of **SPP** matrices. In that M-matrices are closed under positive diagonal equivalence, the same can be said for **IM** matrices.

First, we identify the case in which no path product equalities occur, i.e., all inequalities (5.12.1) are strict, and call such **SPP** matrices **totally strict path product** (**totally SPP**). Observe that totally **SPP** matrices are necessarily positive. Totally **SPP** is not necessary for **EIM** (just consider the **IM** matrix $B = A[\{2,3,4\}]$ of Example 5.16.6), but we will see (Theorem 5.16.11) that totally **SPP** is sufficient for **EIM**.

Recall (Section 5.12) that if (5.12.4) is satisfied by an **SPP** matrix, we say that A is **purely strict path product matrix** (**PSPP**). We will see (Theorem 5.16.7) that, in **PSPP** matrices, path product equalities force certain cofactors to vanish. Notice that condition (5.12.4) does not hold for Example 5.16.5 because $a_{24} = 0.25 = (0.5)(0.5) = a_{23}a_{34}$ while $a_{14} = 0.4 > (0.7)(0.5) = a_{13}a_{34}$ and $a_{21} = 0.5 > (0.5)(0.7) = a_{23}a_{31}$. We show that any **IM** matrix is purely **SPP** (Theorem 5.16.7) and that *EIM* is equivalent to purely **SPP** (Theorem 5.16.12). We note that purely **SPP** and **SPP** coincide (vacuously) when $n \leq 3$ and that generally the **TSPP** matrices are contained in the purely **SPP** matrices (vacuously). Last, but important, observe that, if A is totally (purely) **SPP**, then so is any normalization of A. Thus, in trying to show a totally (purely) **SPP** matrix is **IM**, by positive diagonal equivalence, we may, and do, assume that A is normalized. Thus, the basic **PP**inequalities are $a_{ij}a_{jk} \leq a_{ik}$ for all i, j, k.

Theorem 5.16.7 Any **IM** matrix is purely **SPP**.

Proof Without loss of generality, let $A = [a_{ij}]$ be a normalized **IM** matrix. If $n \leq 3$, then A is purely **SPP** vacuously. So we may assume that $n \geq 4$. Assume that $a_{ik} = a_{ij}a_{jk}$ for the distinct indices i, j, k of N and let $m \in N - \{i, j, k\}$. Consider the principal submatrix

$$A[\{m,j,k,i\}] = \begin{bmatrix} 1 & a_{mj} & a_{mk} & a_{mi} \\ a_{jm} & 1 & a_{jk} & a_{ji} \\ a_{km} & a_{kj} & 1 & a_{ki} \\ a_{im} & a_{ij} & a_{ik} & 1 \end{bmatrix}$$

of A. This submatrix is **IM** by inheritance. So (via, for instance, the special case of Sylvester's identity for determinants given in Section 5.9.1), the (k, i) cofactor $c_{ki} = (a_{mk} - a_{mj}a_{jk})(a_{im} - a_{ij}a_{jm}) \leq 0$. Hence, by condition (5.12.4), either $a_{mk} = a_{mj}a_{jk}$ or $a_{im} = a_{ik}a_{km}$. Thus, A is purely **SPP**. □

A graph may be constructed on the path product equality triples with an edge corresponding to coincidence of the first two or last two indices. This can be informative about the structure of path product equalities, but all we need here is the following lemma. The important idea is that the occurrence of **path product equalities** ensure that certain submatrices have rank one and, moreover, these rank one submatrices are large enough to guarantee that certain almost principal minors vanish. We make use of the well-known fact that an n-by-n matrix is singular if it contains an s-by-t submatrix of rank r such that $s+t \geq n+r+1$.

Lemma 5.16.8 If (i,j,k) is a path product equality for the n-by-n purely **SPP** matrix A, $n \geq 4$, then $\det A(\{k\},\{i\}) = 0$, i.e., the (k,i) cofactor of A vanishes. Moreover, for all real t, $\det A^{(t)}(\{k\},\{i\}) = 0$.

Proof Without loss of generality, assume that A is an n-by-n normalized, purely **SPP** matrix, $n \geq 4$, and that (i,j,k) is a path product equality for A. By permutation similarity, we may assume that $(i,j,k) = (1,2,3)$ so that $a_{13} = a_{12}a_{23}$. Then, because A is purely **SPP**, for each $c \in \langle n\rangle - \{1,2,3\}$, either $a_{1c} = a_{12}a_{2c}$ or $a_{c3} = a_{c2}a_{23}$. Without loss of generality, assume that $\{c \mid c \in \langle n\rangle - \{1,2,3\}$ and $a_{1c} = a_{12}a_{2c}\} \neq \emptyset$. By permutation similarity, we may assume, again without loss of generality, that, for some $q \in \langle n\rangle - \{1,2,3\}$,

(i) $a_{1c} = a_{12}a_{2c}, c = 3, \ldots, q$;

and

(ii) $a_{1c} \neq a_{12}a_{2c}, c = q+1, \ldots, n$.

Note that, for $c \in \{3, \ldots, q\}$ and $r \in \{q+1, \ldots, n\} \subseteq \langle n\rangle - \{1,2,c\}$, (i) implies (by condition (5.12.4)) that either $a_{1r} = a_{12}a_{2r}$ or $a_{rc} = a_{r2}a_{2c}$. Hence, by (ii), we have

(iii) $a_{rc} = a_{r2}a_{2c}, r = q+1, \ldots, n, c = 3, \ldots, q$.

If $q = n$ and $B = A[\{1,2\},\{2, \ldots, q\}]$, then (i) implies that $[a_{12}\, a_{13}\, \ldots\, a_{1q}]$, the first row of B, is a_{12} times $[1\, a_{23}\, a_{24}\, \ldots\, a_{2q}]$, the second row of B. Thus, B is a 2-by-$(n-1)$ rank one submatrix of $A(\{3\},\{1\})$. Because $2+(n-1) = n+1 \geq n+1 = (n-1)+1+1$, $A(\{3\},\{1\})$ is singular.

On the other hand, if $3 \leq q < n$, let $C = A[\{1,2,q+1, \ldots, n\},\{2, \ldots, q\}]$. Then, (i) implies that $[a_{12}\, a_{13}\, \ldots\, a_{1q}]$, the first row of C, is a_{12} times $[1\, a_{23}\, a_{24}\, \ldots\, a_{2q}]$, the second row of C. Now consider $[a_{r2}\, a_{r3}\, \ldots\, a_{rq}]$, the rth row of C, $r = q+1, \ldots, n$. It follows from (iii) that the rth row of C is a_{r2} times the second row of C, $r = q+1, \ldots, n$. Hence,

C is a $(n-q+2)$-by-$(q-1)$ rank one submatrix of $A(\{3\},\{1\})$. Because $(n-q+2)+(q-1)=n+1\geq n+1=(n-1)+1+1$, $A(\{3\},\{1\})$ is singular in this case also. Thus, $\det A(\{3\},\{1\})=0$ in either case, completing the proof of the first part.

The second part follows by the same argument because (5.12.4) holds for A if and only if it holds for $A^{(t)}$ for any real number t. □

Now using Lemma 5.16.8, we have

Corollary 5.16.9 If (i,j,k) is a path product equality for the n-by-n **IM** matrix A, $n\geq 4$, then $(A^{-1})_{ki}=0$.

Theorem 5.16.10 Let A be an n-by-n **SPP** matrix. Then there exists $T>0$ such that $\det A^{(t)}>0$ for all $t>T$.

Proof Without loss of generality, let A be an n-by-n normalized **SPP** matrix, and let $\max_{i\neq j} a_{ij}=M<1$. Denote the set of permutations of $\langle n\rangle$ by S_n and the identity permutation by *id*. Then,

$$\begin{aligned}\det A^{(t)} &= \sum_{\tau\in S_n} sgn(\tau)a^t_{1,\tau(1)}a^t_{2,\tau(2)}\cdots a^t_{n,\tau(n)}\\ &= 1+\sum_{\substack{\tau\in S_n\\ \tau\neq id}} sgn(\tau)a^t_{1,\tau(1)}a^t_{2,\tau(2)}\cdots a^t_{n,\tau(n)}\\ &> 1-\sum_{\substack{\tau\in S_n\\ \tau\neq id}} a^t_{1,\tau(1)}a^t_{2,\tau(2)}\cdots a^t_{n,\tau(n)}\\ &> 1-(n!-1)M^t.\end{aligned}$$

It is clear that there exists $T>0$ such that for all $t>T$, $1-(n!-1)M^t>0$, completing the proof. □

Theorem 5.16.11 Let A be an n-by-n totally **SPP** matrix. Then there exists $T>0$ such that $A^{(t)}\in$ **IM** for all $t>T$.

Proof Without loss of generality, let A be an n-by-n normalized, totally **SPP** matrix. As in the proof of Theorem 5.16.10, let $\max_{i\neq j} a_{ij}=M<1$, let S_n denote the set of permutations of $\langle n\rangle$, and let *id* denote the identity permutation. Then, it follows from Theorem 5.16.10 that there exists $T_1>0$ such that, for all $t>T_1$, $\det A^{(t)}>0$.

Now let $i,j\in\langle n\rangle$ with $i\neq j$ and let $N_1=\langle n\rangle-\{i,j\}$. Consider $c^{(t)}_{ij}$, the i,j cofactor of $A^{(t)}$. Without loss of generality, assume that $i<j$. Then, if $B=\begin{bmatrix} A^{(t)}[N_1] & A^{(t)}[N_1,i]\\ A^{(t)}[j,N_1] & a^t_{ji}\end{bmatrix}$,

$$\begin{aligned}
c_{ij}^{(t)} &= (-1)^{i+j} \det A^{(t)}(\{i,j\}) \\
&= (-1)^{i+j}(-1)^{n-i-1}(-1)^{n-j} \det B \\
&= (-1) \det B \\
&= -\sum_{\tau \in S_{n-1}} sgn(\tau) b_{1,\tau(1)} b_{2,\tau(2)} \dots b_{n-1,\tau(n-1)} \\
&= -a_{ji}^t - \sum_{\substack{\tau \in S_{n-1} \\ \tau \neq id}} sgn(\tau) b_{1,\tau(1)} b_{2,\tau(2)} \dots b_{n-1,\tau(n-1)} \\
&\leq -a_{ji}^t + \sum_{\substack{\tau \in S_{n-1} \\ \tau \neq id}} b_{1,\tau(1)} b_{2,\tau(2)} \dots b_{n-1,\tau(n-1)}.
\end{aligned}$$

Now let

$$\begin{aligned}
\Lambda_1 &= \{\tau \in S_{n-1} \mid \tau \neq id \text{ and } \tau(n-1) = n-1\} \\
&= \left\{\tau \in S_{n-1} \mid \tau \neq id \text{ and } b_{n-1,\tau(n-1)} = a_{ji}^t\right\}
\end{aligned}$$

and

(1) $\Lambda_2 = \{\tau \in S_{n-1} \mid \tau(n-1) \neq n-1\} = \left\{\tau \in S_{n-1} \mid b_{n-1,\tau(n-1)} \neq a_{ji}^t\right\},$

so that $\{id\}$, Λ_1, and Λ_2 form a partition of S_{n-1}. Thus,

$$\begin{aligned}
\sum_{\substack{\tau \in S_{n-1} \\ \tau \neq id}} b_{1,\tau(1)} b_{2,\tau(2)} \dots b_{n-1,\tau(n-1)} &= \sum_{\tau \in \Lambda_1} b_{1,\tau(1)} b_{2,\tau(2)} \dots b_{n-1,\tau(n-1)} \\
&\quad + \sum_{\tau \in \Lambda_2} b_{1,\tau(1)} b_{2,\tau(2)} \dots b_{n-1,\tau(n-1)}.
\end{aligned}$$

Each term of the first summation on the right-hand side of (1) is the product of $b_{n-1,\tau(n-1)} = a_{ji}^t$ and a "non-identity" term of the expansion of $\det A[N_1]$ (because $\tau \neq id$). Hence, each term in this summation has a factor of the form

(2) $$a_{ji}^t a_{si_1}^t a_{i_1 i_2}^t \dots a_{i_{k-1} i_k}^t a_{i_k s}^t$$

in which $s, i_1, \dots, i_k$ are distinct indices in N_1 and $k \geq 1$. Therefore, the cycle product (2) has at least three terms. Because the factors of this term distinct from a_{ji}^t are <1, each term in the first summation is $< a_{ji}^t M^{2p}$ and there are $|S_{n-2}| - 1 = (n-2)! - 1$ such terms.

On the other hand, each term of the second summation on the right-hand side of (1) has a factor of the form

(3) $$a_{ji_1}^t a_{i_1 i_2}^t a_{i_2 i_3}^t \dots a_{i_{k-1} i_k}^t a_{i_k i}^t$$

in which $i_1, \ldots, i_k$ are distinct indices in N_1 with $k \geq 1$. Hence, the path product (3) has at least two terms. Let m_{ji} denote the maximum (j, i) path product given by (3). Therefore, each term in the second summation is $\leq m_{ji}$, and there are $(n-1)! - (n-2)! = (n-2)((n-2)!)$ such terms. Notice that, because A is totally **SPP**, $m_{ji} < a_{ji}$. Thus,

$$\begin{aligned} c_{ij}^{(t)} &\leq -a_{ji}^t + ((n-2)! - 1)\, a_{ji}^t M^{2p} + (n-2)((n-2)!)\, m_{ji}^t \\ &= -a_{ji}^t \left(1 - ((n-2)! - 1)\, M^{2p} - (n-2)((n-2)!) \left(\frac{m_{ji}}{a_{ji}}\right)^t\right). \end{aligned}$$

Because $M < 1$ and $m_{ji} < a_{ji}$, there exists $T_{ij} > 0$ such that, for all $t > T_{ij}$, $c_{ij}^{(t)} \leq 0$. Let $T_2 = max_{i \neq j} T_{ij}$. Then, for all $t > max(T_1, T_2) = T$, the inverse of $A^{(t)}$ has non-positive off-diagonal entries and, hence, $A^{(t)} \in$ **IM**, completing the proof. □

Our main result characterizes eventually inverse M-matrices and, in a certain sense, provides a converse to the statement that **IM** implies (some kind of) **SPP**.

Theorem 5.16.12 Let A be an n-by-n nonnegative matrix. Then, A is purely **SPP** if and only if $A \in$ **EIM**.

Proof We observe that, for either of the two properties in question, purely **SPP** or **EIM**, we may assume that A is an n-by-n normalized **SPP** matrix. Necessity of the condition purely **SPP** then follows from Theorem 5.16.7 and the previously noted fact that the path product inequalities (and equalities) are preserved for any positive Hadamard power t.

For sufficiency, let A be an n-by-n normalized, purely **SPP** matrix. It follows from Theorem 5.16.10 that there exists $T_1 > 0$ such that $\det A^{(t)} > 0$ for all $t > T_1$. So we are left to show that, for some $T > 0$, the off-diagonal entries of $(A^{(t)})^{-1}$ are nonpositive for all $t > T$. To this end, let $i, j \in \langle n \rangle$ with $i \neq j$, and $N_1 = \langle n \rangle - \{i, j\}$. If $a_{ji} = a_{jk}a_{ki}$ for some $k \in N_1$, then it follows from Lemma 5.16.8 that $c_{ij}^{(t)}$, the i, j cofactor of $A^{(t)}$, vanishes for all positive p. Let $T_{ij} = 1$ in this case.

On the other hand, suppose that

$$a_{ji} > a_{jk}a_{ki}$$

for all $k \in N_1$. Following the proof of Theorem 5.16.10, let $\max_{i \neq j} a_{ij} = M < 1$ and let m_{ji} denote the maximum (j, i) path product given by (3). That $m_{ji} < a_{ji}$ follows from (2). Hence, (1) implies that there is a positive constant T_{ij} such that, for all $t > T_{ij}$, $c_{ij}^{(t)}$, the (i, j) cofactor of $A^{(t)}$ is ≤ 0. Letting $T_2 = \max_{i \neq j} T_{ij}$ and $T = \max(T_1, T_2)$, we see that for all $t > T$, $A^{(t)}$ is invertible and its inverse has non-positive off-diagonal entries; that is, $A^{(t)}$ is **IM**. So A is **EIM**, completing the proof. □

Remark 5.16.13 Suppose that we have a totally (purely) **SPP** matrix that is not **IM**. Because the totally (purely) **SPP** matrices are closed under any positive Hadamard power, extraction of a small enough Hadamard root will produce a totally (purely) **SPP** matrix in which P must be arbitrarily large, while raising the matrix to a large enough Hadamard power will produce a totally (purely) **SPP** matrix in which P may be taken to be an arbitrarily small positive number. In fact, T can be 0.

Immediately from Theorem 5.16.10, we have

Corollary 5.16.14 If A is an **IM**-matrix, then there exists $T > 0$ such that $A^{(t)} \in$ **IM** for all $t > T$.

Recall that **TSPP** matrices are necessarily positive (see Section 5.12). While the condition **TSPP** is sufficient for **EIM**, positive **EIM** matrices are not necessarily **TSPP** (see Example 5.12.7). Rather, the correct necessary and sufficient condition, **PSPP** (see Section 5.12), was given in [JS07a] for positive matrices. The general (nonnegative) case follows by the same argument, and we have

Theorem 5.16.15 Let A be a nonnegative n-by-n matrix. A is **EIM** if and only if A is **PSPP**.

Theorem 5.16.16 For an n-by-n nonnegative matrix A, either

(i) there is no $t > 0$ such that $A^{(t)}$ is **IM** or
(ii) there is a critical value $T > 0$ such that $A^{(t)}$ is **IM** for all $t > T$ and $A^{(t)}$ is not **IM** for all $0 \leq t < T$.

The situation for **IM** matrices (which includes the symmetric ones) should be contrasted with **doubly nonnegative matrices (DN)**, i.e., those matrices that are symmetric positive semi-definite and entry-wise nonnegative. Note that a symmetric **IM** matrix is **DN**. For the n-by-n **DN** matrices as a class, there is a **critical exponent** T such that $A \in$ **DN** implies $A^{(t)} \in$ **DN** for all $t \geq T$ and T is a minimum over all **DN** matrices [HJ91]. That critical exponent is $n - 2$ [Hor90, HJ91]. All positive integer Hadamard powers of **DN** matrices are **DN** (because the positive semi-definite matrices are closed under Hadamard product), but it is possible for non-integer powers to leave the class, until the power increases to $n-2$. This, curiously, cannot happen for (symmetric) **IM** matrices, as the "critical exponent" for the entire **IM** class is simply equal to 1.

As mentioned, a positive matrix (certainly) need not be **SPP**, and **SPP** matrices need not be **IM**. However, it is worth noting that the addition of a multiple of the identity can "fix" both of these failures. If $A > 0$ is n-by-n, it is shown in [JS07a] that there is an $\alpha \geq 0$ such that $\alpha I + A$ is **SPP**; in addition, either A is already **SPP** or a value $\beta > 0$ may be calculated such that $\alpha I + A$ is **SPP** for all $\alpha > \beta$. Moreover, if $A \geq 0$ is n-by-n and **SPP**, then there is a minimal $\beta \geq 0$ such that $\alpha I + A$ is **IM** for all $\alpha > \beta$. In fact, if A is normalized **SPP**, which may always be arranged (Section 5.12), then $\beta \leq n - 2$. This means that if we consider $A \circ B$ in which A and B are in **IM**, then either $A \circ B$ will be **IM** (if, for example, one of the matrices is already of the form: a large multiple of I plus an **IM**) or may be made **IM** by the addition of a positive diagonal matrix (that is not too big).

5.17 Perturbation of IM Matrices

Summary. Here we discuss the effect of several types of perturbation and their effect on an **IM** matrix. These perturbations are positive rank one perturbations, positive diagonal perturbations, and perturbations induced by interval matrices. How may a given **IM** matrix be altered so as to remain **IM**, and how may a non-**IM** matrix be changed so as to become **IM**? As mentioned in Theorem 5.5.2, the addition of a nonnegative diagonal matrix to an **IM** matrix results in an **IM** matrix. We add that some nonnegative matrices that are not **IM** may be made **IM** via a nonnegative diagonal addition.

If A is the inverse of an M-matrix that has no zero minors, then each entry (column or row) of A may be changed, at least a little, so as to remain **IM**. By linearity of the determinant, the set of possibilities for a particular entry (column or row) is an interval (convex set), which suggests the question of determination of this interval (convex set).

5.17.1 Positive Rank One Perturbations

We begin by discussing positive rank one perturbation of a given **IM** matrix. There is the following nice result, found in [JS11, theorem 9.1].

Theorem 5.17.1 Let A be an **IM** matrix, let p and q be arbitrary nonnegative vectors, and for $t \geq 0$, define

$$x = Ap,$$
$$y^T = q^T A,$$

and

$$s = 1 + tq^T Ap.$$

We then have

(i) $\left(A + txy^T\right)^{-1} = A^{-1} - \frac{t}{s}pq^T$ is an M-matrix;
(ii) $\left(A + txy^T\right)^{-1} x = \frac{1}{s}p \geq 0$ and $y^T\left(A + txy^T\right)^{-1} = \frac{1}{s}q^T \geq 0$;
(iii) $y^T\left(A + txy^T\right)^{-1} x = \frac{1}{s}q^T Ap < \frac{1}{t}, t > 0.$

This perturbation result may be used to show very simply that so-called **strictly ultrametric** matrices [Fie98a] are **IM** (see Section 5.28).

If we consider a particular column of an **IM** matrix, then the set of replacements of that column that result in an **IM** matrix is a convex set. This convex set may be viewed as the intersection of $n^2 - n + 2$ half-spaces [JS11, theorem 1.2]. Without loss of generality, we may assume the column is the last, so that, partitioned by columns,

$$A(x) = [a_1\ a_2\ \ldots\ a_{n-1}\ x],$$

with $a_1, a_2, \ldots, a_{n-1} \geq 0$. Then the half-spaces are given by the linear constraints

$$x \geq 0,$$
$$(-1)^{i+j+1} \det A(x)(i,j) \geq 0,\ \ 1 \leq i \leq n,\ 1 \leq j < n,\ i \neq j,$$

and

$$\det A(x) > 0.$$

A similar analysis may be given for a single off-diagonal entry. It may be taken to be the $(1,n)$ entry, so that

$$A(x) = \begin{bmatrix} a_{11} & x \\ A_{21} & A_{22} \end{bmatrix}$$

with x a scalar. Now the interval for x is determined by the inequalities

$$(-1)^{i+j+1} \det A(x)(i,j) \geq 0,\ \ 1 < i \leq n,\ 1 \leq j < n,\ i \neq j,$$

and

$$\det A(x) > 0.$$

In [JS11] conditions are given on

$$A = [a_1\ a_2\ \ldots\ a_{n-1}]$$

such that there exist an $x \geq 0$ so that

$$A(x) = [a_1\ a_2\ \ldots\ a_{n-1}\ x]$$

is **IM**. If A is re-partitioned as

$$A = \begin{bmatrix} A_{11} \\ a_{21} \end{bmatrix}$$

in which A_{11} is square, the conditions are the following subset of those in [JS11, theorem 2.4]:

(i) A_{11} is **IM**;
(ii) $a_{21} \geq 0$;
(iii) $a_{21}A_{11}^{-1} \geq 0$.

5.17.2 Positive Diagonal Perturbations

What about diagonal perturbation? By Theorem 5.5.2, if $D \in \overline{\mathcal{D}}$, then $A + D$ is **IM** whenever A is. So what if A is not **IM**? If A is irreducible, we need to assume that $A > 0$. Is anything else necessary for $A + D$ to be **IM** for some $D \in \overline{\mathcal{D}}$? Interestingly, we will see that an $A > 0$ may always be made **PP** and **SPP** by a sufficiently large diagonal addition.

Necessary and sufficient conditions for a nonnegative matrix to be **IM** are as follows.

Lemma 5.17.2 If A is an n-by-n nonnegative matrix, then A is **IM** if and only if

(i) its determinant and its maximal proper principal minors are positive and
(ii) its maximal almost principal minors are signed as those of an inverse M-matrix.

Here we show that a positive matrix may be made **PP** by well-defined additions to the diagonal. We establish a determinantal inequality for normalized **SPP** matrices. For normalized **IM** matrices, this inequality provides an interesting dominance relation between the principal minors and certain associated *APM*s. We then show that the addition of the identity matrix to any 4-by-4 normalized **SPP**-matrix results in an **IM** matrix. This 4-by-4 fact may be generalized; we show that any n-by-n **SPP** matrix can be made **IM** by the addition of any scalar matrix sI such that $s \geq n - 3$. (In fact, there is a lower bound on

s that is less than or equal to $n-3$.) The latter result has implications for pairs of **IM** matrices whose Hadamard product is **IM**.

5.17.3 Positive, Path Product, and Inverse M-Matrices

Our first two propositions clarify the connection between positive matrices and positive **PP** matrices.

Proposition 5.17.3 Let $A = [a_{ij}]$ be an n-by-n positive matrix and suppose that [JS01a, (1.1)] holds for all distinct i, j, k. Then,

$$a_{ij}a_{ji} \leq a_{ii}a_{jj}.$$

Proof Suppose that $i, j \in \langle n \rangle$ with $i \neq j$. Then, for all $k \neq i, j$, we have

$$a_{ij}a_{ji} \leq \left(\frac{a_{ii}a_{kj}}{a_{ki}}\right)\left(\frac{a_{jj}a_{ki}}{a_{kj}}\right) = a_{ii}a_{jj},$$

because $a_{ki}a_{ij} \leq a_{kj}a_{ii}$ and $a_{kj}a_{ji} \leq a_{jj}a_{ki}$. □

We note that Proposition 5.17.3 shows that, as in prior work [JS07b], the distinctness of the indices need not be required in the definition of **PP**.

Proposition 5.17.4 Let $A = [a_{ij}]$ be an n-by-n matrix such that $a_{ij} > 0$, $i \neq j$, and $a_{ii} = 0$, $i = 1, \ldots, n$. Then, there is a unique, minimum positive diagonal matrix D such that $A + D$ is **PP**.

Proof Let $D = \text{diag}(d_1, \ldots, d_n)$ in which d_k is given by

$$d_k = \max_{\substack{i \neq j \\ i,j \neq k}} \frac{a_{ik}a_{kj}}{a_{ij}}.$$

This assures that $a_{ij} \geq \frac{a_{ik}a_{kj}}{d_k}$ for triples $\{i, j, k\}$ that are distinct and according to [JS07a, proposition 1] those for which $i \neq j$, but $i = k$, follow. □

A direct consequence of Proposition 5.17.4 is that the addition of a (minimal) nonnegative diagonal matrix or a minimal positive scalar matrix converts a square positive matrix into a **PP** matrix. A priori this diagonal matrix is not bounded, even in terms of n, but it is bounded in terms of the sizes and ratios among entries. We also note that, depending on the zero pattern of the entries, we may not be able to convert a nonnegative matrix with zero diagonal to a **PP** matrix with positive diagonal matrix addition. For instance, if $A = [a_{ij}]$

is such a matrix and for distinct indices i, j, k, $a_{ij}a_{jk} > 0$ while $a_{ik} = 0$, then $A + D \notin PP$ for any positive diagonal matrix D.

We next establish a determinantal inequality for certain normalized **SPP** matrices. For normalized **IM** matrices, the inequality shows that the dominance relationship between diagonal and associated off-diagonal entries extends to all proper principal minors and certain associated *APM*s. Observe that a special case of this inequality is the known fact that a row (column) diagonally dominant M matrix has an inverse that is diagonally dominant of its column (row) entries. The theorem establishes the following useful fact: a normalized **SPP** matrix is **IM** provided that the proper principal minors and the *APM*s are appropriately signed. (In this case, the determinant is positive.) The following theorem and corollary were crucial to the proof of the main result of [Che07a].

Theorem 5.17.5 Let $A = [a_{ij}]$ be an n-by-n normalized **SPP** matrix, $n \geq 2$, whose proper principal minors are positive and whose *APM*s are signed as those of an **IM**-matrix. Then,

(i) for any nonempty proper subset α of $\langle n\rangle = \{1, 2, \ldots, n\}$ and for any indices $i \in \alpha$ and $j \notin \alpha$

$$\det A[\alpha] > \max\{|\det A[\alpha - i + j, \alpha]|, |\det A[\alpha, \alpha - i + j]|\}; \quad (5.17.1)$$

(ii) $\det A > 0$; and

(iii) A is **IM**.

Proof Because the theorem obviously holds for $n = 2$, we assume henceforth that $n \geq 3$. To establish (5.17.1), it suffices to prove that

$$\det A[\langle n-1\rangle] > |\det A[\langle n-1\rangle, \langle n\rangle - 1]|.$$

First, suppose that $a_{1n} = 0$, and let $S = \{i \mid a_{1i} = 0\}$, k denote the cardinality of S, and $T = \{i \mid a_{1i} \neq 0, i \in \langle n\rangle - 1\}$. If $k = n - 1$, then [JS07a, (1.5)] obviously holds. So we may assume that S and T are each nonempty and form a partition of $\langle n\rangle - 1$. Let $i \in T$ and $j \in S$. Then, by the PP conditions, $a_{1i}a_{ij} \leq a_{1j} = 0$. Because $a_{1i} \neq 0$, this implies $a_{ij} = 0$ for $i \in T$ and $j \in S$, and, in addition, $a_{1j} = 0$ for $j \in S$, hence, $A[\langle n-1\rangle, \langle n\rangle - 1]$ has an $(n-k)$-by-k zero submatrix. It follows from the Frobenius–König Theorem that $\det A[\langle n-1\rangle, \langle n\rangle - 1] = 0$, which completes the proof of this case.

Now suppose that $a_{1n} > 0$ and n is odd (the even case is analogous). Because the almost principal minors of A are signed as those of an **IM** matrix, it follows that the $n, 1$ cofactor of A, $(-1)^{n+1} \det A[\langle n-1\rangle, \langle n\rangle - 1] \leq 0$, and, hence, $\det A[\langle n-1\rangle, \langle n\rangle - 1] \leq 0$ (because n is odd). If $\det A[\langle n-1\rangle, \langle n\rangle - 1] = 0$, then the result certainly holds. So assume that $\det A[\langle n-1\rangle, \langle n\rangle - 1] < 0$. Then,

$$
\begin{aligned}
|\det A[\langle n-1\rangle, \langle n\rangle - 1]| &= -\det A[\langle n-1\rangle, \langle n\rangle - 1] \\
&= -\sum_{i=1}^{n-1} (-1)^{i+n-1} a_{in} \det A[\langle n-1\rangle - i, \langle n-1\rangle - 1] \\
&\quad \text{(expanding by the last column)} \\
&= \sum_{i=1}^{n-1} (-1)^{i+1} a_{in} \det A[\langle n-1\rangle - i, \langle n-1\rangle - 1] \\
&= a_{1n} \{\det A[\langle n-1\rangle - 1] \\
&\quad + \sum_{i=2}^{n-1} (-1)^{i+1} \frac{a_{in}}{a_{1n}} \det A[\langle n-1\rangle - i, \langle n-1\rangle - 1]\} \\
&< \det A[\langle n-1\rangle - 1] \\
&\quad + \sum_{i=2}^{n-1} (-1)^{i+1} \frac{a_{in}}{a_{1n}} \det A[\langle n-1\rangle - i, \langle n-1\rangle - 1] \\
&\quad \text{(because } 0 < a_{1n} < 1\text{)} \\
&\le |\det A[\langle n-1\rangle - 1]| \\
&\quad + \sum_{i=2}^{n-1} (-1)^{i+1} a_{i1} \det A[\langle n-1\rangle - i, \langle n-1\rangle - 1] \\
&\quad \left(\begin{array}{l}\text{because each term in the summand has the same} \\ \text{sign as the } i,1 \text{ cofactor of } A[\langle n-1\rangle] \text{ and hence is} \\ \text{nonpositive, and because } \dfrac{a_{in}}{a_{1n}} \ge a_{i1}\end{array}\right) \\
&= \det A[\langle n-1\rangle],
\end{aligned}
$$

which establishes (5.17.1). Applying (5.9.5) and (5.17.1), we see that (b) (and hence (c)) holds, completing the proof. □

Then, directly from Theorem 5.17.5, we have

Corollary 5.17.6 Let A be an n-by-n nonnegative matrix, $n \ge 2$, each of whose proper principal submatrices is **IM**. Then A is **IM** if and only if each of its maximal *APM*s is signed as that of an **IM** matrix.

We also have the following dominance relation between the principal minors and certain associated *APM*s of a normalized **IM** matrix.

Corollary 5.17.7 Let $A = [a_{ij}]$ be an n-by-n normalized **IM** matrix, $n \geq 2$. Then, for α a nonempty proper subset of $\langle n \rangle = \{1, 2, \ldots, n\}$ and for any indices $i \in \alpha$ and $j \notin \alpha$, (5.17.1) holds.

Examining the proof of Theorem 5.17.5 carefully, we see that we did not need all proper principal minors and all *APM*s to be appropriately signed for the matrix to have positive determinant, i.e., we have

Proposition 5.17.8 Let A be an n-by-n nonnegative matrix, $n \geq 3$. Then, A is **IM** if and only if A is **SPP** and all principal minors and all *APM*s, each of orders $n-1$ and $n-2$, are signed as those of an **IM** matrix.

The following theorem, based primarily on path product inequalities, motivated this work, and we will see that it is a special case of more general results that follow.

Theorem 5.17.9 Let A be a 4-by-4 normalized **SPP** matrix. Then $A+I$ is **IM**. Furthermore, $A + sI$, $s < 1$, need not be **IM**.

Proof By Theorem 5.17.5 and permutation similarity, to show $A + I$ is **IM**, it suffices to show that a particular 3-by-3 *APM* is necessarily properly signed. We show that the (4, 1) *APM* is nonnegative:

$$\det(A+I)(4,1) = \det \begin{bmatrix} a_{12} & a_{13} & a_{14} \\ 1+1 & a_{23} & a_{24} \\ a_{32} & 1+1 & a_{34} \end{bmatrix}$$

$$= 4a_{14} + a_{12}a_{23}a_{34} + a_{13}a_{32}a_{24} - 2a_{12}a_{24} - 2a_{13}a_{34} - a_{14}a_{23}a_{32}$$

$$= 2(a_{14} - a_{12}a_{24} + a_{14} - a_{13}a_{34}) + a_{12}a_{23}a_{34} + a_{13}a_{32}a_{24} - a_{14}a_{23}a_{32}$$

$$\geq 2(a_{14} - a_{12}a_{24} + a_{14} - a_{13}a_{34})$$

$$+ a_{12}a_{23}a_{32}a_{34} + a_{13}a_{32}a_{23}a_{34} - a_{14}a_{23}a_{32}$$

(by PP) If the sum of the last three terms is nonnegative, then the determinant is nonnegative by the path product inequalities. Otherwise,

$$= 2(a_{14} - a_{12}a_{24} + a_{14} - a_{13}a_{34}) + (a_{12}a_{34} + a_{13}a_{34} - a_{14})a_{23}a_{32}$$

$$\geq 2(a_{14} - a_{12}a_{24} + a_{14} - a_{13}a_{34}) + (a_{12}a_{34} + a_{13}a_{34} - a_{14})$$

$$= 2(a_{14} - a_{12}a_{24}) + (a_{14} - a_{13}a_{34}) + a_{12}a_{34}$$

$$\geq 0.$$

For the second claim, let $0 < \epsilon$, $s < 1$ and let

$$A = \begin{bmatrix} 1 & 1-\epsilon & 1-\epsilon & 1-\epsilon \\ \epsilon & 1 & \epsilon & 1-\epsilon \\ \epsilon & \epsilon & 1 & 1-\epsilon \\ \epsilon & \epsilon & \epsilon & 1 \end{bmatrix}.$$

Then $A+sI$ is **SPP**. Further, for ϵ "small," $\det(A+sI)(4,1) \approx (s+1)(s-1) < 0$ so that $A+sI$ is not **IM**, completing the proof. □

Let $A = [a_{ij}]$ be an n-by-n normalized **SPP** matrix, $n \geq 3$, and, for $i \neq j$, let

$$u_{ij}^{(A)} = \begin{cases} \dfrac{1}{a_{ij}} \displaystyle\sum_{\substack{k=1 \\ k \neq i,j}}^{n} a_{ik}a_{kj}, & a_{ij} \neq 0 \\ 0, & a_{ij} = 0. \end{cases}$$

(Notice that $a_{ij} = 0$ implies $a_{ik}a_{kj} = 0$ for all k.) Let $U(A) = \max_{i \neq j} u_{ij}^{(A)}$. We call $U(A)$ the **upper path product bound** for A.

The following straightforward lemma will prove useful.

Lemma 5.17.10 Let $A = [a_{ij}]$ be an n-by-n normalized **SPP** matrix, $n \geq 3$. Then,

(i) for any $i, j \in N$ with $i \neq j$,

$$\sum_{\substack{k=1 \\ k \neq i,j}}^{n} a_{ik}a_{kj} \leq U(A)a_{ij};$$

(ii) $U(A) \leq n-2$. Moreover, $U(A) = n-2$ if and only if, for some $i, j \in N$ $(i \neq j)$, $a_{ij} \neq 0$ and $a_{ik}a_{kj} = a_{ij}$ for all $k \neq i, j$.

Proof The conclusions follow from the definition of $U(A)$ and [JS07a, (1.1)]. □

The next two theorems generalize Theorem 5.17.9.

Theorem 5.17.11 Let $A = [a_{ij}]$ be an n-by-n normalized **SPP** matrix, $n \geq 3$, and let $L = \max\{U(A), 1\}$. Then $A + sI$ is **IM** for all $s \geq L - 1$. Furthermore, $L - 1$ may not be replaced by a smaller value.

Proof By induction on n. If $n = 3$, then $U(A) \leq 1$ so that $L = 1$. Because A is **SPP**, $A = A + 0I = A + (L-1)I$ is **IM** and thus $A + sI$ is **IM** for all $L \geq L_1$, where establishing the first claim for $n = 3$. Proceeding inductively, it follows that the $(n-1)$-by-$(n-1)$ principal minors of $B = A + (L-1)I$

are positive because for any principal submatrix A_1 of A, $A_1 + (L_1 - 1)I$ is **IM** so that $A_1 + (L - 1)I$ is **IM**, as $L \geq L_1$, where $L_1 = \max\{U(A_1), 1\}$. From Theorem 5.17.5, it follows that, if the entries of adj B are properly signed, then $\det B > 0$ and B is **IM**. Thus, for the first claim, it suffices to verify that an $(n-1)$-by-$(n-1)$ *APM* is properly signed. By permutation similarity, it suffices to check any particular one, say the complement of the $(1,2)$-entry, which should be nonnegative. Its value is

$$b_{21} \det B(\{1,2\}) - \begin{bmatrix} b_{23} & \dots & b_{2n} \end{bmatrix} adj\, B(\{1,2\}) \begin{bmatrix} b_{31} \\ \vdots \\ b_{n1} \end{bmatrix}.$$

Division by $\det B(\{1,2\})$ means that we need

$$b_{21} \geq \begin{bmatrix} b_{23} & \dots & b_{2n} \end{bmatrix} B(\{1,2\})^{-1} \begin{bmatrix} b_{31} \\ \vdots \\ b_{n1} \end{bmatrix}. \tag{5.17.2}$$

Let $B(\{1,2\})^{-1}$ have entries c_{ij}, $i,j = 3, \dots, n$. By induction, $C = [c_{ij}]$ is an M-matrix. The right-hand side of (5.17.2) is

$$\sum_{i,j=3}^{n} b_{2i}c_{ij}b_{j1} = \sum_{i \neq j} b_{2i}c_{ij}b_{j1} + \sum_{i=3}^{n} b_{2i}c_{ii}b_{i1}. \tag{5.17.3}$$

Because $c_{ij} \leq 0$, $i \neq j$, the first term on the right-hand side of (5.17.3) is not more than $\sum_{i \neq j} b_{2i}c_{ij}b_{ji}b_{i1}$ by path product.

Because of Fischer's inequality [HJ91] applied to the **IM** matrix $B(\{1,2\})$, we have $\det B(\{1,2\}) \leq b_{ii} \det B(\{1,2,i\}) = L \det B(\{1,2,i\})$ so that $\frac{1}{L} \leq \frac{\det B(\{1,2,i\})}{\det B(\{1,2\})} = c_{ii}$.

Combining, we obtain

$$\begin{aligned}
\sum_{i=3}^{n}\sum_{j=3}^{n} b_{2i}c_{ij}b_{j1} &= \sum_{i=3}^{n}\sum_{\substack{j=3 \\ j\neq i}}^{n} b_{2i}c_{ij}b_{j1} + \sum_{i=3}^{n}(b_{2i}c_{ii}b_{i1} + b_{2i}c_{ii}b_{ii}b_{i1} - b_{2i}c_{ii}b_{ii}b_{i1}) \\
&\leq \sum_{i=3}^{n}\sum_{j=3}^{n} b_{2i}c_{ij}b_{ji}b_{i1} + \sum_{i=3}^{n}(1 - b_{ii})b_{2i}c_{ii}b_{i1} \\
&\qquad (\text{because } b_{j1} = a_{j1} \geq a_{ji}a_{i1} = b_{ji}b_{i1} \geq 0 \text{ and } c_{ij} \leq 0, i \neq j) \\
&= \sum_{i=3}^{n} b_{2i}b_{i1} \sum_{j=3}^{n} c_{ij}b_{ji} + \sum_{i=3}^{n}(1 - L)b_{2i}c_{ii}b_{i1}
\end{aligned}$$

$$
\begin{aligned}
&= \sum_{i=3}^{n} b_{2i} b_{i1} (1 + (1 - L) c_{ii}) \\
&\left(\text{because} \sum_{j=3}^{n} c_{ij} b_{ji} = \text{the i, i entry of } CC^{-1} \right) \\
&\leq \sum_{i=3}^{n} b_{2i} b_{i1} \left(1 + (1 - L) \frac{1}{L} \right) \\
&= \frac{1}{L} \sum_{i=3}^{n} b_{2i} b_{i1} \\
&= \frac{1}{L} \sum_{i=3}^{n} a_{2i} a_{i1} \\
&\leq \frac{1}{L} (U(A)\, a_{21}) \\
&\leq a_{21},
\end{aligned}
$$

which completes the proof of the first claim.

If the normalized **SPP** matrix

$$
A = \begin{bmatrix}
1 & 1-\epsilon & 1-\epsilon & \dots & 1-\epsilon & 1-\epsilon \\
\epsilon & 1 & \epsilon & \dots & \epsilon & 1-\epsilon \\
\epsilon & \epsilon & \ddots & \ddots & \vdots & \vdots \\
\vdots & \vdots & \ddots & \ddots & \epsilon & 1-\epsilon \\
\vdots & \vdots & & \ddots & 1 & 1-\epsilon \\
\epsilon & \epsilon & \dots & \dots & \epsilon & 1
\end{bmatrix}
$$

in which $0 < \epsilon < 1$ and $C = [c_{ij}]$ is the cofactor matrix of $B = A + sI$, then, for ϵ "small,"

$$
\begin{aligned}
c_{n1} &\approx (-1)^{n+1}(-1)^{n}((1+s)^{n-2} - (n-2)(1+s)^{n-3}) \\
&= -(1+s)^{n-3}(s - (n-3)).
\end{aligned}
$$

So, if $0 < s < L - 1 \leq n - 3$, $\det B > 0$ and $c_{n1} > 0$. Hence, the $1, n$ entry of B^{-1} is positive. Thus, B is not **IM**, establishing the second claim. □

From Lemma 5.17.10 (ii), we have

Theorem 5.17.12 Let A be an n-by-n normalized **SPP** matrix, $n \geq 3$. Then $A + sI$ is **IM** for all $s \geq n - 3$.

A consequence of Theorem 5.17.5 is the following.

Corollary 5.17.13 Let A be an n-by-n nonnegative matrix with positive diagonal entries, and let D and E be positive diagonal matrices such that $DE = (n-2)[\text{diag}(A)]^{-1}$. Then, if $DAE - (n-3)I$ is **SPP**, A is **IM**.

Theorem 5.17.14 Let $A > 0$ be given and let D and E be positive diagonal matrices such that $DE = (n-2)[\text{diag}(A)]^{-1}$. Then, if $DAE - (n-3)I$ is **PP**, $A \in$ **IM**.

Theorem 5.17.15 Let A be an n-by-n normalized **SPP** matrix, $n \geq 3$. Then, there is a minimal real number $s_0(A)$ such that for all $s \geq s_0(A)$, $A + sI$ is **IM**.

Proof Let $A = [a_{ij}]$ be a normalized **SPP** matrix, s be a real number, and $c_{ij}(s)$ denote the (i,j) cofactor of $A + sI$, $i \neq j$. Then,

$$\begin{aligned} c_{ij}(s) &= (-1)^{i+j} \det (A + sI)(j,i) \\ &= -\det \begin{bmatrix} a_{ij} & A[i,\sigma] \\ A[\sigma,j] & (A+sI)[\sigma] \end{bmatrix} \\ &= -a_{ij}(1+s)^{n-2} + \text{lower order terms in } (1+s) \end{aligned}$$

in which $\sigma = \langle n \rangle - \{i,j\}$. Because $c_{ij}(s) \to -\infty$ as $s \to \infty$, there is a smallest real number $s_{ij}(A)$ such that $c_{ij}(s) \leq 0$ for all $s \geq s_{ij}(A)$. Let $s_0(A) = \max_{i \neq j} s_{ij}(A)$. Then, by Theorem 5.17.12, $A + sI$ is **IM** for all $s \geq s_0(A)$. □

Example 5.17.16 Consider the 4-by-4 normalized **SPP** matrix

$$A = \begin{bmatrix} 1 & 0.10 & 0.40 & 0.30 \\ 0.40 & 1 & 0.40 & 0.65 \\ 0.10 & 0.20 & 1 & 0.60 \\ 0.15 & 0.30 & 0.60 & 1 \end{bmatrix}.$$

As seen in Section 5.2, A is not **IM** (the $(2,3)$-entry of A^{-1} is positive). By actual calculation, $U(A) = \frac{1}{a_{31}}(a_{32}a_{21} + a_{34}a_{41}) = 1.7 > 1$. Hence, $A + sI$ is **IM** for all $s \geq 0.7$. $\left(\text{In fact, } A + sI \text{is } \mathbf{IM} \text{ if and only if } s \geq \frac{\sqrt{1537}-25}{80} \approx 0.18.\right)$

If $\mathcal{C}$ is a class of matrices, the *Hadamard dual* of $\mathcal{C}$, denoted IM^D, is the set of matrices $A \in \mathcal{C}$ such that $A \in SC$ if and only if $A \circ B \in \mathcal{C}$ for all $B \in \mathcal{C}$. In [HJ91] an example was given to show that the Hadamard product of two independent inverse M-matrices need not be **IM**. Later a 4-by-4 symmetric example was given [WZZ00]. Because the Hadamard product of two inverse M-matrices is inverse M when $n \leq 3$ (Section 5.16), this fully clarifies when the class of **IM** matrices is itself contained its Hadamard dual. Obviously, the positive diagonal matrices and J, the all ones matrix, lie in the Hadamard dual of the n-by-n **IM** matrices for any positive integer n. Because **IM** matrices are closed under positive diagonal multiplication, the latter implies that the positive rank one matrices also lie in the dual. Beyond this, the question is: do the n-by-n **IM** matrices, $n \geq 4$, have anything else in their Hadamard dual? It is noted in Section 5.2 that, if $A = [a_{ij}]$ is an n-by-n **IM** matrix, then there exist positive diagonal matrices D and E such that $A = DA_1E$ in which $A_1 = [\alpha_{ij}]$ is a normalized inverse M-matrix. Because $D(A \circ B)E = A \circ (DBE)$ for positive diagonal matrices D and E, we only need to test A with normalized **IM** matrices B to see if A is in the Hadamard dual.

Observe that $(DAE) \circ (FBG) = DF(A \circ B)GE$ for positive diagonal matrices D, E, G, and H and that $A \circ B$ is normalized **SPP** provided A and B are

Lemma 5.17.17 If A is an n-by-n normalized **IM** matrix, then $(A+(n-3)I_n) \in$ **IM**^{D}.

Proof Let A be an n-by-n normalized **IM** matrix and B be an n-by-n **IM** matrix. Then, there exist positive diagonal matrices D and E such that $B = DB_1E$ in which B_1 is a normalized **IM** matrix. Thus, $(A + (n-3)I_n) \circ B = D[A \circ B_1 + (n-3)I_n]E$ and the result follows from the remarks above. □

Theorem 5.17.18 If A is an n-by-n **IM** matrix and D and E are positive diagonal matrices such that $A_1 = DAE$ is normalized, then $A + (n-3)D^{-1}E^{-1} \in$ **IM**^{D}.

Proof It follows from Lemma 5.17.20 that $A_1 + (n-3)I_n \in$ **IM**^{D}. Hence, $A + (n-3)D^{-1}E^{-1} = D^{-1}(A_1 + (n-3)I_n)E^{-1} \in$ **IM**^{D}. The result follows from Theorem 5.17.12. □

As we noted previously, the Hadamard product of two **IM** matrices need not be **IM** for $n > 3$. However, because of Lemma 5.17.20, the Hadamard product of any **IM** matrix with many others is **IM**. Theorem 5.17.18 identifies a number of matrices in **IM**^{D}, and any **IM**-matrix may be changed to one in **IM**^{D} by addition of an appropriate scalar matrix.

5.17.4 Intervals of IM Matrices

Our last perturbation result pertains to **IM** interval matrices. If $A, B \in M_n(\mathbb{R})$, the **interval from A to B**, denoted by $I(A, B)$ is the set of matrices $C = [c_{ij}] \in M_n(\mathbb{R})$ satisfying $\min\{a_{ij}, b_{ij}\} \leq c_{ij} \leq \max\{a_{ij}, b_{ij}\}$ for all i and j, while the set of **vertices (vertex matrices)** derived from A and B, denoted $V(A, B)$, is the set of matrices $C = [c_{ij}] \in M_n(\mathbb{R})$ such that $c_{ij} = a_{ij}$ or b_{ij} for all i and j. If $C = [c_{ij}]$ in which $c_{ij} = min\{a_{ij}, b_{ij}\}$ ($c_{ij} = max\{a_{ij}, b_{ij}\}$), $i,j = 1, \ldots, n$, then C is called the **left endpoint matrix** (**right endpoint matrix**). Note that there are at most 2^{n^2} distinct vertex matrices. We were motivated by the following question raised by J. Garloff [Gar]: given the interval determined by two **IM** matrices, when are all matrices in the interval **IM**? We fully answer this question by showing that all matrices in the interval are **IM** if and only if $V(A,B) \subseteq$ **IM**. Then, by example, we show that there are limitations upon strengthening the answer and note that the same technique generalizes our answer to any **identically signed class** (**IS**) (a class of matrices defined by the signs weakly or strongly of a certain collection of minors.) The signs may be weak or strong. One might suspect that the interval between any two **IM** matrices is contained in the class of **IM** matrices if the two matrices are comparable with respect to the usual entry-wise partial ordering. An example is given to show that this conjecture is false (although it does hold for M-matrices [Fan60]). See Section 5.10 for a discussion of almost principal minors of **M** and **IM** matrices.

Recall that a totally positive (nonnegative) matrix is a matrix with all minors positive (nonnegative). We denote the totally positive (nonnegative) matrices by **TP** (**TN**). Thus, **M** (**IM**, **TP**, **TN**) matrices, all have identically signed principal and almost principal minors. Of course, some of the defining inequalities may be weak inequalities and others strong. (For example, the almost principal minor inequalities for an M-matrix are weak, while the principal minor inequalities are strong.) Note that each of the classes mentioned so far is an **IS** class.

Example 5.17.19 Consider the **IM** matrices

$$\mathbf{A} = \begin{bmatrix} 1 & .4 & .3 \\ .6 & 1 & .6 \\ .4 & .6 & 1 \end{bmatrix} \leq \mathbf{B} = \begin{bmatrix} 1 & .9 & .6 \\ .6 & 1 & .6 \\ .4 & .6 & 1 \end{bmatrix}.$$

The matrix $\mathbf{C} = \begin{bmatrix} 1 & .6 & .3 \\ .6 & 1 & .6 \\ .4 & .6 & 1 \end{bmatrix}$ satisfies $A \leq C \leq B$ (entry-wise), yet C is not an **IM** matrix because $(C^{-1})_{13} > 0$.

The conclusion of Example 5.17.19 remains true even if we restrict the set of matrices under consideration to the matrices whose inverses are tridiagonal M-matrices (under the entry-wise partial ordering).

A **line in a matrix** is a single row or column.

Lemma 5.17.20 Suppose that $A = [a_{ij}]$ and $B = [b_{ij}] \in \mathbf{IM}$, and that $a_{ij} = b_{ij}$ except perhaps for the entries in one line. Then, for $0 \le t \le 1$, $tA + (1-t)B \in \mathbf{IM}$.

Proof Because A and B are both **IM**, all their corresponding principal and almost principal minors agree in sign (perhaps weakly in the almost principal case). Because the only entries that vary lie in, at most, a single line, each minor of $tA+(1-t)B$ is a linear function of t. Thus, each principal or almost principal minor of $tA + (1-t)B$, $0 \le t \le 1$, agrees in sign (perhaps weakly in the almost principal case) with the corresponding minor of A (or B). Because all key minors are correctly signed, this implies that $tA + (1-t)B$, $0 \le t \le 1$, is an **IM** matrix that completes the proof. □

An immediate consequence of Lemma 5.17.20 is the following fact that we will use.

Corollary 5.17.21 Suppose that $A = [a_{ij}]$ and $B = [b_{ij}] \in \mathbf{IM}$, and that $a_{ij} = b_{ij}$ for $(i,j) \neq (r,s)$, i.e., A and B differ in at most the r,s entry. Then, $tA + (1-t)B \in \mathbf{IM}$ for $0 \le t \le 1$.

In the event that A and B differ in only one entry, the interval determined by A and B is the same as the line segment determined by them. If they differ in more entries (even in the same line), then the interval is a larger set than the line segment because independent variation in each entry is allowed.

Moreover, the lemma does not necessarily hold if two matrices differ in more than a line.

Example 5.17.22 Consider the **IM** matrices

$$\mathbf{A} = \begin{bmatrix} 1 & 0 & 0 \\ .6 & 1 & .6 \\ .4 & .6 & 1 \end{bmatrix} \le \mathbf{B} = \begin{bmatrix} 1 & .6 & 0 \\ .6 & 1 & 0 \\ .4 & .6 & 1 \end{bmatrix}.$$

For $0 < t < 1$,

$$tA + (1-t)B = \begin{bmatrix} 1 & .6(1-t) & 0 \\ .6 & 1 & .6t \\ .4 & .6 & 1 \end{bmatrix}$$

cannot possibly be **IM** because it is irreducible, but contains a zero entry (see [Joh82]).

Theorem 5.17.23 Let $A, B \in M_n(\mathbb{R})$. Then $V(A,B) \subseteq \mathbf{IM}$ if and only if $I(A,B) \subseteq \mathbf{IM}$.

Proof First assume that $V(A,B) \subseteq \mathbf{IM}$. Then A and B are both in **IM**. Let $C = [c_{ij}] \in I(A,B)$. If we replace the $(1,1)$ entry of each matrix in $V_0 = V(A,B)$ with c_{11} to obtain a new set of matrices V_1, it follows from the corollary that $V_1 \subseteq \mathbf{IM}$. Similarly, if we replace the $(1,2)$ entry of each matrix in V_1 with c_{12} to obtain the set of matrices V_2, then $V_2 \subseteq \mathbf{IM}$. Proceeding in this manner, filling in the matrices left to right by rows with entries from C, we replace the (i,j) entry of each matrix in $V_{n(i-1)+j-1}$ with c_{ij} and the resulting set of matrices $V_{n(i-1)+j} \subseteq \mathbf{IM}$ and have half as many elements as $V_{n(i-1)+j-1}$. In the last step of this process, we obtain the set of matrices V_{n^2} whose sole element is C. Because $V_{n^2} \subseteq \mathbf{IM}$, C is an **IM** matrix. Thus, $I(A,B) \subseteq \mathbf{IM}$, and because the converse is obvious, this completes the proof. □

Note that, in regard to Theorem 5.17.18, $V(A,B)$ consists of A, B, and $\begin{bmatrix} 1 & .4 & .6 \\ .6 & 1 & .6 \\ .4 & .6 & 1 \end{bmatrix} \in \mathbf{IM}$, and $\begin{bmatrix} 1 & .9 & .3 \\ .6 & 1 & .6 \\ .4 & .6 & 1 \end{bmatrix} \notin \mathbf{IM}$. Thus, we see that we cannot expect the statement of the theorem to hold for the entry-wise partial order if "all vertices" is replaced by all but one of the vertices. Moreover, as the following example shows, we cannot expect the statement to hold for the "checkerboard" partial ordering, i.e.,

$$A \preccurlyeq B \Leftrightarrow (-1)^{i+k} a_{ik} \leq (-1)^{i+k} b_{ik}, i,k = 1, \ldots, n,$$

either, even if the two endpoints of the interval (the "corner matrices" [Gar96]) are **IM**. Notice that this latter order corresponds to the sign pattern of the inverse of a *TP*-matrix.

Example 5.17.24 Consider the **IM** matrices

$$\mathbf{A} = \begin{bmatrix} 1 & .9 & .6 \\ .6 & 1 & .6 \\ .4 & .6 & 1 \end{bmatrix} \preccurlyeq \mathbf{B} = \begin{bmatrix} 1 & .4 & .6 \\ .6 & 1 & .6 \\ .4 & .3 & 1 \end{bmatrix}.$$

The matrix $\mathbf{C} = \begin{bmatrix} 1 & .9 & .6 \\ .6 & 1 & .6 \\ .4 & .3 & 1 \end{bmatrix}$ satisfies $A \preccurlyeq C \preccurlyeq B$, yet C is not an **IM** matrix because $(C^{-1})_{32} > 0$.

Last, we consider the partial order $A \preccurlyeq B$, which corresponds to the sign pattern of the inverse of an **IM** matrix, namely

$$A \preccurlyeq B \Leftrightarrow -a_{ik} \leq -b_{ik}, i \neq k; a_{ik} \leq b_{ik}, i = k, i,k = 1, \ldots, n,$$

then Theorem 5.17.23 again shows that we cannot expect the statement to hold even if the two corner matrices are **IM**.

In light of this example, it is not obvious that there exists a partial order for which it suffices to check only the corners or any subset of the vertices as was the case for *TP*-matrices [Gar82] (in which case it was only necessary to check the two corner matrices) and invertible *TN*-matrices [Gar96] (in which case it was only necessary to test a maximum of 2^{2n-1} vertices).

We close by noting that the proof of Lemma 5.17.20 shows that for A, B that lie in any particular **IS** class and differ in at most one line, the line segment joining them lies fully in the same **IS** class. Using the corresponding corollary, a corresponding theorem is that for any particular **IS** class, it is necessary (obviously) and sufficient, for the entire interval to be in that **IS** class, that the vertices be in that **IS** class.

5.18 Determinantal Inequalities

Classical determinantal inequalities associated with the names of Hadamard, Fischer, Koteljanskii, and Szasz have long been known for M-matrices. Because of Jacobi's determinantal identity, it is an easy exercise to show that these also hold for **IM** matrices. The most general of these, associated with the name Koteljanskii, is

Theorem 5.18.1 If J, K are index sets contained in $\langle n \rangle$ and $A \in \mathbf{IM}$ is n-by-n, then

$$\det A[J \cup K] \det A[J \cap K] \leq \det A[J] \det A[K].$$

Proof As in the work of Koteljanskii, this may be proven using Sylvester's determinantal inequality, permutation similarity invariance of **IM** and the fact that symmetrically placed almost principal minors of $A \in \mathbf{IM}$ are weakly of the same sign. □

The inequalities of Hadamard, Fischer, and Szasz may be deduced from Theorem 5.18.1 and are stated below.

Corollary 5.18.2 (Hadamard) If $A \in \mathbf{IM}$, then $\det A \leq \prod_{i=1}^{n} a_{ii}$.

Corollary 5.18.3 (Fischer) If $A \in \mathbf{IM}$ and $J \subseteq \langle n \rangle$, then $\det A \leq \det A[J] \det A[J^c]$.

Corollary 5.18.4 (Szasz) If $A \in \mathbf{IM}$ and Π_k denotes the product of all the $\binom{n}{k}$ principal minors of A of size k-by-k, then

$$\Pi_1 \geq (\Pi_2)^{\frac{1}{\binom{n-1}{1}}} \geq (\Pi_3)^{\frac{1}{\binom{n-1}{2}}} \geq \cdots \geq \Pi_n$$

in which $\binom{n}{i}$ denotes the i-th binomial coefficient, $i = 0, 1, \ldots, n$.

There are inequalities among products of principal minors of any $A \in \mathbf{IM}$ besides those in Theorem 5.18.1 and its chordal generalizations. Recently, all such inequalities have been characterized [FHJ98]. In order to understand this result, consider two collections $\alpha = \{\alpha_1, \ldots, \alpha_p\}$ and $\beta = \{\beta_1, \ldots, \beta_p\}$ of index sets, $\alpha_i, \beta_j \subseteq \langle n \rangle$, $i, j \in \{1, \ldots, p\}$. For any index set $J \subseteq \langle n \rangle$ and any collection α, define the two functions

$$f_\alpha(J) \equiv \text{ the number of sets } \alpha_i \text{ such that } J \subseteq \alpha_i$$

and

$$F_\alpha(J) \equiv \text{ the number of sets } \alpha_i \text{ such that } \alpha_i \subseteq J.$$

For the two collections α, β, the following two set-theoretic axioms are important:

(ST0) $$f_\alpha(\{i\}) = f_\beta(\{i\}),\ i = 1, \ldots, n$$

and

(ST2) $$F_\alpha(J) \geq F_\beta(J), \text{ for all } J \subseteq N.$$

A third axiom (ST1) arises only in the characterization of determinantal inequalities for M-matrices. The result for **IM** is then

Theorem 5.18.5 The following statements about two collections α, β of index sets are equivalent:

(i) $\frac{\prod_{i=1}^t \det A[\alpha_i]}{\prod_{i=1}^t \det A[\beta_i]}$ is bounded over all $A \in$ **IM**;
(ii) $\prod_{i=1}^t \det A[\alpha_i] \leq \prod_{i=1}^t \det A[\beta_i]$ for all $A \in$ **IM**; and
(iii) the pair of collections α, β satisfy ST0 and ST2.

The proof is given in [FHJ98].

The above results leave only the question of whether there are inequalities involving some non-principal minors in a matrix $A \in$ **IM**. Because **IM** matrices are P-matrices, there are some obvious inequalities, e.g.,

$$\det A[\alpha + i, \alpha + j] \det A[\alpha + j, \alpha + i] \leq \det A[\alpha + i] \det A[\alpha + j]$$

whenever $i, j \notin \alpha$. There is also a family of nontrivial inequalities involving almost principal minors of **IM** matrices that extend those of Theorem 5.11.1. These inequalities exhibit a form of monotonicity already known for principal minors. Recall that if $\alpha \subseteq \beta \subseteq \langle n \rangle$, then

$$\det A[\beta] \leq \det A[\alpha] \left(\prod_{k \in \beta - \alpha} a_{kk} \right) \text{ for } A \in \textbf{IM}.$$

This just follows from det $A[\beta] \le$ det $A[\alpha]$ det $A[\beta - \alpha]$ and Hadamard's inequality (both are special cases of Theorem 5.19.1). If A were normalized, the product of diagonal entries associated with $\beta - \alpha$ would disappear and the above inequality could be paraphrased "bigger minors are smaller." The same holds for our new inequalities, which generalize those given in Theorem 5.11.1.

Theorem 5.18.6 Let $\alpha \subseteq \beta \subseteq \langle n \rangle$, and suppose that $A = [a_{ij}] \in$ **IM** is n-by-n. Then, if $\det A[\alpha + i, \alpha + j] \neq 0$,

$$\frac{|detA[\beta + i, \beta + j]|}{|detA[\alpha + i, \alpha + j]|} \le \frac{\det A[\beta \cup \{i,j\}]}{\det A[\alpha \cup \{i,j\}]} \le \det\, A[\beta - \alpha] \le \prod_{i \in \beta - \alpha} a_{ii}$$

whenever $i \neq j,\ i,j \notin \beta$. If det $A[\alpha + i, \alpha + j] = 0$, then det $A[\beta + i, \beta + j] = 0$, also, for $i \neq j,\ i,j \notin \beta$.

We note that other determinantal inequalities for **IM** matrices were given in [Che07b].

5.19 Completion Theory

A **partial matrix** is an array with some entries specified, and the other, *unspecified*, entries free to be chosen. A **completion of a partial matrix** is the conventional matrix resulting from a particular choice of values for the unspecified entries. [Joh90] is a good reference on matrix completion problems.

One topic of interest is the completion of **partial PP (SPP) matrices**, i.e., nonnegative matrices such that every specified path satisfies the **PP** (**SPP**) conditions in [JS01a, theorem 4.1]. We make the assumption throughout that all diagonal entries are 1s because **PP** matrices are invariant under positive diagonal scaling.

The **SPP** matrix completion problem is fundamental in considering the (difficult) **IM** matrix completion problem [JS96]. Here, a **partial IM-matrix** is a partial nonnegative matrix in which each fully specified principal submatrix is **IM**, and we wish to determine whether a given partial **IM** matrix can be completed to **IM**. Because every **IM** matrix is **SPP**, for such a completion to exist it is necessary for the partial **IM** matrix to be partial **SPP** and for any **IM** matrix completion to be **SPP**. Thus, the set of **SPP** completions (if they exist) of a partial **IM** matrix is a place to start in the search for an **IM** matrix completion, and it represents a narrowing of the superset of possible completions. From [JS01a], we have

Theorem 5.19.1 Every partial **PP** (**SPP**) matrix has a **PP** (**SPP**) matrix completion.

Partial **IM** matrices are not necessarily partial **PP** as shown by the following example.

Example 5.19.2 Consider the partial **IM** matrix

$$\mathbf{A} = \begin{bmatrix} 1 & \frac{1}{2} & ? & \frac{1}{9} \\ \frac{1}{2} & 1 & \frac{1}{2} & ? \\ ? & \frac{1}{2} & 1 & \frac{1}{2} \\ \frac{1}{9} & ? & \frac{1}{2} & 1 \end{bmatrix}.$$

A is not partial **PP** because $a_{12}a_{23}a_{34} = \frac{1}{8} > \frac{1}{9} = a_{14}$. Thus, A cannot be completed to a **PP** matrix and, because every **IM** matrix is **PP**, no **IM** matrix completion exists.

In fact, even if a partial **IM** matrix is partial **SPP**, it may not have an **IM** matrix completion. For instance, if in Example 5.12.7, we let the $(1, 4)$ and $(4, 1)$ entries be unspecified, it can be shown [JS96] that no **IM** matrix completion exists.

A chordal graph is **k-chordal** [Gol80] if no two distinct maximal cliques intersect in more than k vertices. In [JS96] the **symmetric IM** (**SIM**) completion problem was studied, and it was shown that, for partial **SIM** matrices, 1-chordal graphs guarantee **SIM** completion.

Theorem 5.19.3 Let G be a 1-chordal graph on n vertices. Then every n-by-n partial **SIM** matrix A, the graph of whose specified entries is G, has a **SIM** completion. Moreover, there is a unique **SIM** completion A_1 of A whose inverse entries are 0 in every unspecified position of A, and A_1 is the unique determinant maximizing **SIM** completion of A.

However, this is not true for chordal graphs in general. Consider the partial **SIM** matrix (which has a 3-chordal graph with two maximal cliques)

$$\mathbf{A} = \begin{bmatrix} 1 & \frac{9}{40} & \frac{1}{5} & \frac{2}{5} & x \\ \frac{9}{40} & 1 & \frac{3}{10} & \frac{1}{2} & \frac{3}{8} \\ \frac{1}{5} & \frac{3}{10} & 1 & \frac{1}{5} & \frac{3}{4} \\ \frac{2}{5} & \frac{1}{2} & \frac{1}{5} & 1 & \frac{1}{4} \\ x & \frac{3}{8} & \frac{3}{4} & \frac{1}{4} & 1 \end{bmatrix}.$$

In [JS99], using (5.9.11) and the cofactor form of the inverse, it was shown that no value of x yields a **SIM** completion.

Cycle conditions are derived that are necessary for the general **IM** completion problem. Further, these conditions are shown to be necessary and sufficient for completability of a partial symmetric **IM** matrix, the graph of whose specified entries is a cycle.

Theorem 5.19.4 Let

$$\mathbf{A} = \begin{bmatrix} 1 & a_1 & & & & & a_n \\ a_1 & 1 & a_2 & & & ? & \\ & a_2 & \cdot & & & & \\ & & \cdot & \cdot & \cdot & & \\ & & & \cdot & \cdot & \cdot & \\ & ? & & & \cdot & \cdot & a_{n-1} \\ a_n & & & & & a_{n-1} & 1 \end{bmatrix}$$

be an n-by-n partial **SIM** matrix, $n \geq 4$. Then A has a **SIM** completion if and only if the cycle conditions

$$\prod_{j \neq i} a_j \leq a_i, i = 1, \dots, n$$

are satisfied

A **block graph** is a graph built from cliques and (simple) cycles as follows: starting with a clique or cycle, sequentially articulate the "next" clique or simple cycle at no more than one vertex of the current graph. A completability criterion for partial **SIM** matrices with block graphs is as follows.

Theorem 5.19.5 Let A be an n-by-n partial **SIM** matrix, the graph of whose specified entries is a block graph. Then A has a **SIM** completion if and only if all minimal cycles in G satisfy the cycle conditions.

In addition, other graphs for which partial **SIM** matrices have **SIM** completions are discussed.

5.20 Connections with Other Matrix Classes

Let $A \geq 0$. A is **totally nonnegative (totally positive)** if all minors of all orders of A are nonnegative (positive); further, A is **oscillatory** if A is totally nonnegative and a power of A is totally positive [Gan59]. In [Mark72] the class of totally nonnegative matrices whose inverses are M-matrices is characterized as follows

Theorem 5.20.1 Suppose A is an invertible, totally nonnegative matrix. Then A^{-1} is an M-matrix if and only if $\det A(i,j) = 0$ for $i + j = 2k$, in which k is a positive integer, and $i \neq j$.

A special class of oscillatory matrices that are **IM** is also investigated: A **Jacobi matrix** is a square matrix $A = [a_{ij}]$ with real entries satisfying $a_{ij} = 0$ for $|i - j| > 1$. Further, A is a **normal Jacobi matrix** if $a_{i,i+1}, a_{i+1,i} < 0$ for $i = 1, 2, \ldots, n - 1$ [Lew80]. In [Lew89] the results in [Mark72] are extended as follows.

Theorem 5.20.2 Let A be a square matrix. Consider the following three conditions.

(i) A^{-1} is totally nonnegative.
(ii) A is an M-matrix.
(iii) A is a Jacobi matrix.

Then any two of the three conditions imply the third.

Theorem 5.20.3 Let A be a square matrix. Consider the following three conditions.

(i) A^{-1} is oscillatory.
(ii) A is an M-matrix.
(iii) A is a normal Jacobi matrix.

Then any two of the three conditions imply the third.

Theorem 5.20.4 Let A be an M-matrix. Then A^{-1} is totally nonnegative if and only if A is a Jacobi matrix, and A^{-1} is oscillatory if and only if A is a normal Jacobi matrix.

In [Pen95] the class of totally nonnegative matrices that are **IM** are fully characterized as follows.

Theorem 5.20.5 Let A be an invertible totally nonnegative n-by-n matrix. Then the following properties are equivalent.

(i) A^{-1} is an M-matrix.
(ii) $\det A(i,j) = 0$ if $|i - j| = 2$.
(iii) A^{-1} is a tridiagonal matrix.
(iv) For any $k \in \{1, 2, \ldots, n - 1\}$, $\det A([i_1, \ldots, i_k], [j_1, \ldots, j_k]) = 0$ if $|i_l - j_l| > 1$ for some $l \in \{1, \ldots, k\}$.

We have the following characterization of **IM** matrices that are totally positive [Pen95].

Theorem 5.20.6 Let A be an invertible n-by-n M-matrix. Then the following properties are equivalent.

(i) A^{-1} is a totally positive matrix.
(ii) A is a tridiagonal matrix.

5.21 Graph Theoretic Characterizations of IM Matrices

In [LN80] graph theoretic characterizations of $(0,1)$-matrices that are **IM** matrices were obtained. A matrix $A \in \mathbb{R}^{n\times n}$ is **essentially triangular** if for some permutation matrix P, $P^{-1}AP$ is a triangular matrix. Let A be an n-by-n matrix and $G(A) = (V(A) = N, E(A))$ be its associated (adjacency) graph, i.e., $(i,j) \in E(A)$ if and only if $a_{ij} \neq 0$, where $V(A)$ is its set of vertices and $E(A)$ is its set of edges. A path from vertex i to vertex j is called an **(i,j)-path**; a path of length k, k being the number of directed edges in the path, is called a **k-path**; and a path of length k from i to j is called an $(i,j|k)$-path. Let H denote the **join of the (disjoint) graphs** G_1 and G_2 in which G_1 consists of a single vertex and G_2 consists of a 2-path. (H is called a **buckle**.) Then,

Theorem 5.21.1 An essentially triangular $(0,1)$-matrix A is **IM** if and only if the associated graph is a partial order and, for every i,j,k, if an $(i,j|k)$-path in $G(A)$ is a maximal (i,j)-path, then it is a unique k-path.

Theorem 5.21.2 A $(0,1)$-matrix A is **IM** if and only if $G(A)$ induces a partial order on its vertices that contains no buckles as induced subgraphs.

In [KB85] we have the following result that relates the zero pattern of a transitive, invertible $(0,1)$ matrix and that of its inverse.

Theorem 5.21.3 Suppose that $A = [a_{ij}]$ is an n-by-n transitive, invertible $(0,1)$ matrix with $B = [b_{ij}] = A^{-1}$. If $i \neq j$ and $a_{ij} = 0$, then $b_{ij} = 0$.

For a matrix B, let $B_{\mathbf{IM}}$ denote the set

$$B_{\mathbf{IM}} = \{\alpha \in \mathbb{R} \colon \alpha I + B \in \mathbf{IM}\}.$$

$B_{\mathbf{IM}}$ is a (possibly empty) ray on the real line, bounded from below, and is nonempty if and only if B is nonnegative and transitive. Then,

Theorem 5.21.4 Let B be an n-by-n nonnegative transitive matrix. Then the ray $B_{\mathbf{IM}}$ is closed, i.e., $B_{\mathbf{IM}} = [\alpha_0, \infty)$ for some $\alpha_0 \geq 0$, if and only if there exists $\beta_0 \in \mathbb{R}$ such that

(i) $B + \beta_0 I \notin \mathbf{IM}$,
(ii) $B + \beta_0 I$ is positive stable and so are its principal submatrices of order $n-1$.

The authors express the infimum of $B_{\mathbf{IM}}$ as a maximal root of a polynomial that depends on the matrix B, differentiating between the cases in which $B_{\mathbf{IM}}$ is open or closed.

In [Lew89] the following generalization was obtained. Consider a nonnegative square matrix $A = [a_{ij}]$ and its associated adjacency graph $G(A) = (V(A), E(A))$. The **weight of an edge** (i,j) of $G(A)$ is then a_{ij}. The **weight of a directed path** is the product of the weights of its edges. The *weight* of a set of paths is the sum of the weights of its paths. For essentially triangular **IM** matrices, the following result was proved.

Theorem 5.21.5 Let A be an essentially triangular nonnegative invertible matrix, and let A_1 be its normalized matrix. Then $A_1 \in \mathbf{IM}$ if and only if $G(A_1)$ is a partial order and the weight of the collection of even paths between any two vertices in $G(A)$ does not exceed the weight of its collection of odd paths.

A graph-theoretic characterization of general **IM** matrices was then given.

Theorem 5.21.6 A nonnegative square matrix A is **IM** if and only if

(i) For every two distinct vertices i and j in $G(A)$, the sum of the even (i,j)-paths does not exceed that of the odd (i,j)-paths.
(ii) All principal minors of A are positive.

A special class of inverse **IM** matrices was also characterized.

5.22 Linear Interpolation Problems

The **linear interpolation problem** for a class of matrices $\mathcal{C}$ asks for which vectors x, y there exists a matrix $A \in \mathcal{C}$ such that $Ax = y$. In [JS01c] the linear interpolation problem is solved for several important classes of matrices, one of which is $\mathcal{M}$ (and hence it is solved for **IM** also). In addition, a transformational characterization is given for M-matrices that refines the known such characterization for P-matrices. These results were extended to other classes of matrices in [Smi01].

5.23 Newton Matrices

For an n-by-n real matrix A with eigenvalues $\lambda_1, \lambda_2, \ldots, \lambda_n$ let

$$S_k = \sum_{i_1 < \ldots < i_k} \lambda_{i_1} \ldots \lambda_{i_k}$$

and $c_k = \frac{1}{\binom{n}{k}} S_k$, with $c_0 = 1$. The matrix A is a **Newton matrix** [JMP09] if $c_k^2 \geq c_{k-1}c_{k+1}$, $k = 1, \ldots, n-1$. If each $c_k > 0$, A is called p-**Newton**. Using the immanantal results of [JJP98], it was observed in [Hol05] that M-matrices are p-Newton. Because p-Newton matrices are closed under inversion [JMP09], it follows that **IM** matrices are also p-Newton.

5.24 Perron Complements of IM Matrices

In [Mey89a, Mey89b] the notion of **Perron complements** was introduced. Specifically, for an n-by-n nonnegative irreducible matrix A, $\beta \subset \langle n \rangle$, and $\alpha = \langle n \rangle - \beta$, the **Perron complement** of $A[\beta]$ in A is defined to be

$$\mathcal{P}(A/A[\beta]) = A[\alpha] + A[\alpha,\beta](\rho(A)I - A[\beta])^{-1}A[\beta,\alpha].$$

Among other things, it was shown that $\rho(\mathcal{P}(A/A[\beta])) = \rho(A)$ and that if A is row stochastic, then so is $\mathcal{P}(A/A[\beta])$. These results were applied to obtain an algorithm for computing the stationary distribution vector for a Markov chain.

In [JX93] the following question was investigated: when are Perron complements primitive or just irreducible? Their answer settled some questions posed in [Mey89a, Mey89b].

In [Nab00] it is established that if A is an irreducible **IM** matrix, then its Perron complements are also **IM**. In fact,

Theorem 5.24.1 Let A be an irreducible **IM** matrix, $\beta \subset \langle n \rangle$, and $\alpha = \langle n \rangle - \beta$. Then, for any $t \in [\rho(A), \infty)$, the matrix

$$\mathcal{P}_t(A/A[\beta]) = A[\alpha] - A[\alpha,\beta](tI - A[\beta])^{-1}A[\beta,\alpha]$$

is invertible and is an **IM** matrix. In particular, the Perron complement $\mathcal{P}(A/A[\beta])$ $(= \mathcal{P}_1(A/A[\beta])$ is **IM**.

Also, for irreducible **IM** matrices whose inverses are tridiagonal, the following result was proved.

Theorem 5.24.2 Let $A \in$ **IM** with A^{-1} irreducible and tridiagonal. Then, for any subsets $\beta \subset \langle n \rangle$ and $\alpha = \langle n \rangle - \beta$, the Perron complement

$$\mathcal{P}(A/A[\beta]) = A[\alpha] + A[\alpha,\beta](I - A[\beta])^{-1}A[\beta,\alpha]$$

is an **IM** matrix whose inverse is irreducible and tridiagonal and hence is (also) totally nonnegative, i.e., all minors are nonnegative.

Further, it was shown that for an n-by-n **IM** matrix A, the inverse of associated principal submatrices of A are sandwiched between the inverses of the Perron complements of A and the inverses of the corresponding Schur complements of A. Last, directed graphs of inverses of Perron complements of irreducible **IM** matrices were investigated.

5.25 Products of IM Matrices

Let $A \in M_n(\mathbb{R})$. In [FMN83] the following question is investigated: when is the matrix A a product of M-matrices (**IM** matrices)? Let Π denote the set of **IM** matrices that are finite products of **IM** matrices. They prove that $A \in \Pi$ if and only if $A = (L_1U_1)(L_2U_2)\dots(L_kU_k)$ in which L_i (U_i) is a lower (upper) triangular **IM** matrix, $i = 1, \dots, n$ if and only if A is a product of elementary matrices that are **IM**.

Let $\mathcal{R}(A) = \frac{\det A}{\prod_{i=1}^{n} a_{ii}}$ and let $\Pi_{\mathcal{R}}$ denote the set of $A \in$ **IM** with $\mathcal{R}(A) = 1$. A necessary and sufficient combinatorial condition is given for a matrix to be in $\Pi_{\mathcal{R}}$. Toeplitz **IM** matrices are also studied.

More detailed analysis of products of **IM** matrices and of M-matrices, from a different perspective, is found in [JOvdD03] which followed on [JOvdD01].

5.26 Topological Closure of IM Matrices

We say that a matrix A belongs to the (topological) *closure* of the **IM** matrices if A is the limit of a convergent sequence of **IM** matrices. If so, we write $A \in \overline{\mathbf{IM}}$. The following theorem was proved in [FJM87].

Theorem 5.26.1 Let A be a p-by-p **IM** matrix, and let Q be a p-by-n nonnegative matrix with exactly one nonzero entry in each column. Then

$$Q^T AQ + D$$

is an **IM** matrix for any n-by-n positive diagonal matrix D.

Several important facts about **IM** matrices follow as special cases.

The following characterization of $\overline{\mathbf{IM}}$ was obtained in [FJM87].

Theorem 5.26.2 Suppose A is a nonnegative n-by-n matrix. Then the following statements are equivalent:

(i) $A \in \overline{\mathbf{IM}}$,
(ii) $(A + D)^{-1} \leq D^{-1}$ for each positive diagonal matrix D,

(iii) $(A+D)^{-1}$ belongs to **M** for each positive diagonal matrix D,
(iv) $(A+\alpha I)^{-1}$ belongs to **M** for all $\alpha > 0$,
(v) $(A+D)^{-1}A \geq 0$ for each positive diagonal matrix D,
(vi) $(I+cA)^{-1} \leq I$ for all $c > 0$, and
(vii) $cA^2(I+cA)^{-1} \leq A$ for all $c > 0$.

The theorem allows the authors to characterize nilpotent matrices on the boundary of $\overline{\mathbf{IM}}$.

Denote by $\mathcal{L}_{rn}$ the set of all nonnegative r-by-n matrices, $r \leq n$, which contain exactly one nonzero entry in each column. If the dimensions are not specified, write simply $\mathcal{L}$. In [FM88b] a matrix A belonging to $\overline{\mathbf{IM}}$ was shown to have an explicit form. Specifically, they prove

Theorem 5.26.3 An n-by-n matrix A is in $\overline{\mathbf{IM}}$ if and only if there exists a permutation matrix P, a diagonal matrix D with positive diagonal entries, a matrix $B \in \mathbf{IM}$, and a matrix $Q \in \mathcal{L}$ without a zero row such that

$$\mathbf{D}^{-1}PAP^TD = \begin{bmatrix} 0 & UBQ & UBV+W \\ 0 & Q^TBQ & Q^TBV \\ 0 & 0 & 0 \end{bmatrix}$$

for some nonnegative matrices U, V, and W.

In the partitioning, any one or two of the three block rows (and their corresponding block columns) can be void.

The preceding theorem allows the characterization of singular matrices in $\overline{\mathbf{IM}}$.

If A is a square matrix, the smallest integer k for which rank (A^k) = rank (A^{k+1}) is called the **index of** A, denoted by index$(A) = k$. The **Drazin inverse** of A is the unique matrix A^D such that

(i) $A^{k+1}A^D = A^k$;
(ii) $A^DAA^D = A^D$;
(iii) $AA^D = A^DA$.

In [FM90] the Drazin inverse of a matrix belonging to $\overline{\mathbf{IM}}$ is determined and, for such a matrix A, the arrangement of the nonzero entries of the powers $A^2, A^3, \ldots$ is shown to be invariant.

5.27 Tridiagonal, Triangular, and Reducible IM Matrices

In [Ima83] **IM** matrices whose nonzero entries have particular patterns were studied. First, tridiagonal **IM** matrices were characterized as follows.

Theorem 5.27.1 Let A be a nonnegative, invertible, tridiagonal n-by-n matrix. Then the following statements are equivalent:

(i) A is an **IM** matrix.
(ii) All principal minors of A are positive, and A is the direct sum of matrices of the following types: (a) diagonal matrices, (b) 2-by-2 positive matrices, or (c) matrices of the form

$$A = \begin{bmatrix} d_1 & a_1 & 0 & \cdot & \cdot & \cdot & \cdot & \cdot & \cdot & \cdot & 0 \\ 0 & d_2 & 0 & 0 & & & & & & & \cdot \\ 0 & b_1 & d_3 & a_2 & 0 & & & & & & \cdot \\ \cdot & 0 & 0 & d_4 & 0 & 0 & & & & & \cdot \\ \cdot & & 0 & b_2 & d_5 & a_3 & 0 & & & & \cdot \\ \cdot & & & 0 & 0 & d_6 & 0 & \cdot & & & \cdot \\ \cdot & & & & \cdot & \cdot & \cdot & \cdot & \cdot & & \cdot \\ \cdot & & & & & \cdot & \cdot & \cdot & \cdot & \cdot & \cdot \\ \cdot & & & & & & \cdot & \cdot & \cdot & \cdot & 0 \\ \cdot & & & & & & & \cdot & \cdot & \cdot & a_t \\ 0 & \cdot & \cdot & \cdot & \cdot & \cdot & \cdot & \cdot & 0 & b_u & d_s \end{bmatrix}$$

where $a_t = 0$ and $u = \frac{s-1}{s}$ when s is odd, and $b_u = 0$ and $t = \frac{s}{2}$ when s is even.

A nonempty subset K of $\mathbb{R}^n$, which is closed under addition and nonnegative scalar multiplication, is called a **(convex) cone** in $\mathbb{R}^n$. If a cone K is also closed topologically, has a nonempty interior, and satisfies $K \cap (-K) = \emptyset$, then K is called a **proper cone**. If A is a real matrix, then the set $K(A) = \{Ax \colon x \geq 0\}$ is a **polyhedral cone**, i.e., a set of nonnegative linear combinations of a finite set S of vectors in $\mathbb{R}^n$. Using these notions, the following geometric characterization of upper triangular **IM** matrices was given [Ima83].

Theorem 5.27.2 Let A be a nonnegative upper triangular n-by-n matrix with 1s along the diagonal. Then the following statements are equivalent.

(i) A is an **IM** matrix.
(ii) $Ae_k - e_k \in K(Ae_1, Ae_2, \dots, Ae_{k-1})$ for $k = 2, \dots, n$.

For certain types of reducible matrices, both necessary conditions and sufficient conditions for the reducible matrix to be **IM** are provided in [Ima84].

5.28 Ultrametric Matrices

$A = [a_{ij}]$ is a **strictly ultrametric matrix** [MMS94] if A is real symmetric and (entry-wise) nonnegative and satisfies

(i) $a_{ik} \geq \min(a_{ij}, a_{jk})$ for all i, j, k; and
(ii) $a_{ii} > \max_{k \neq i} a_{ik}$ for all i.

In [MMS94] it was shown that if $A = [a_{ij}]$ is strictly ultrametric, then A is invertible and $A^{-1} = [\alpha_{ij}]$ is a strictly diagonally dominant Stieltjes matrix satisfying $\alpha_{ij} = 0$ if and only if $a_{ij} = 0$. The proof utilized tools from topology and real analysis. Then, a simpler proof that relied on tools from linear algebra was given in [NV94]. Such statements, and some following, also follow easily from [JS11, theorem 9.1], giving a yet simpler proof.

In [Fie98b] new characterizations of matrices whose inverse is a weakly diagonally dominant symmetric M-matrix are obtained. The result in [MMS94] is shown to follow as a corollary. In [Fie98b] connections between ultrametric matrices and Euclidean point spaces are explored. See also [Fie76].

There has been a great deal of work devoted to generalizations of ultrametric matrices. In [NV95] and [MNST95], nonsymmetric ultrametric (called **generalized ultrametric**) matrices were independently defined and characterized. Necessary and sufficient conditions are given for regularity, and in the case of regularity, the inverse is shown to be a row and column diagonally dominant M-matrix. In [MNST96] nonnegative matrices whose inverses are M-matrices with **unipathic graphs** (a digraph is **unipathic** if there is at most one simple path from any vertex i to any vertex k) are characterized. A symmetric ultrametric matrix A is **special (symmetric ultrametric matrix)** [Fie00] if, for all i,

$$a_{ii} = \max_{k \neq i} a_{ik}.$$

In [Fie00] graphical characterizations of special ultrametric matrices are obtained. In [MSZ03], by using dyadic trees, a new class of nonnegative matrices is introduced, and it is shown that their inverses are column diagonally dominant M-matrices.

In [Stu98] a polynomial time spectral decomposition algorithm to determine whether a specified, real symmetric matrix is a member of several closely related classes is obtained. One of these classes is the class of strictly ultrametric matrices.

In [TV99] it is shown that a classical inequality due to Hardy can be interpreted in terms of symmetric ultrametric matrices, and then a generalization of this inequality can be derived for a particular case.

6

Copositive Matrices

6.1 Introduction

Let $\Delta_n = \left\{x \in \mathbb{R}^n : x \geq 0 \text{ and } e^T x = 1\right\}$ be the **unit simplex** of $\mathbb{R}^n$. For convenience in this chapter, we let $M_n(\{-1, 0, 1\})$ and $M_n(\{-1, 1\})$ denote the n-by-n matrices whose entries lie in the indicated sets.

Definition 6.1.1 A symmetric matrix $A \in M_n(\mathbb{R})$ is

(i) **Copositive** ($A \in \mathbf{C}$) if $x \in \Delta_n$ implies $x^T Ax \geq 0$;
(ii) **Strictly copositive** ($A \in \mathbf{SC}$) if $x \in \Delta_n$ implies $x^T Ax > 0$;
(iii) **Copositive-plus** ($A \in \mathbf{C}^+$) if A is copositive and $x \in \Delta_n, x^T Ax = 0$ imply $Ax = 0$.

Of course, the requirement $x \in \Delta_n$ may be replaced by $x \in \mathbb{R}^n$, $0 \neq x \geq 0$. Note that $\mathbf{SC} \subseteq \mathbf{C}^+ \subseteq \mathbf{C}$.

The classes **C** and **SC** are broader than the positive semidefinite (**PSD**) and positive definite (**PD**) classes, respectively, as requirements are placed upon fewer vectors. If **sN** denotes the symmetric, nonnegative matrices in $M_n(\mathbb{R})$, then both **PSD**, $\mathbf{sN} \subseteq \mathbf{C}$, so that $\mathbf{PSD} + \mathbf{sN} \subseteq \mathbf{C}$.

Copositive matrices were first defined by Motzkin in 1952 [Mot52]. Much of the early, formative work on copositivity ([Mot52, Dia62, Bau66, Bau67, Bas68]) focused on whether $\mathbf{C} = \mathbf{PSD} + \mathbf{sN}$ for general n (Diananda's conjecture, [Dia62]). For $n = 2$, $\mathbf{C} = \mathbf{PSD} \cup \mathbf{sN}$, and $\mathbf{C} = \mathbf{PSD} + \mathbf{sN}$ for $n = 3, 4$ [Dia62]. However, the question was settled negatively by an example of Alfred Horn, reported in [Hal67].

Example 6.1.2 Let $A \in M_5(\mathbb{R})$ be the symmetric circulant

$$A = \begin{bmatrix} 1 & -1 & 1 & 1 & -1 \\ -1 & 1 & -1 & 1 & 1 \\ 1 & -1 & 1 & -1 & 1 \\ 1 & 1 & -1 & 1 & -1 \\ -1 & 1 & 1 & -1 & 1 \end{bmatrix}.$$

Then $A \in \mathbf{C}$ (see the general result in [Hal67]), but A is not the sum of a **PSD** matrix and a nonnegative matrix, as any symmetric decrease in a pair of symmetrically placed off-diagonal entries, or in a diagonal entry, leaves a matrix that is not **PSD**. The matrix itself has two negative eigenvalues and is not **PSD**. Counterexamples to the Diananda conjecture for $n > 5$ may be constructed by direct summation, so that $\mathbf{C} \subseteq \mathbf{PSD} + \mathbf{sN}$ if and only if $n \leq 4$.

A copositive matrix not in $\mathbf{PSD} + \mathbf{sN}$ is called an **exceptional (copositive) matrix**. The copositive matrices form a convex cone in $M_n(\mathbb{R})$, and the quadratic form $x^T Ax$ of a matrix on an extremal of this cone is called an **extreme copositive quadratic form**.

The important Example 6.1.2 has only entries in $\{-1, 1\}$. This has brought a lot of attention to $\mathbf{C} \cap M_n(\{-1, 1\})$ and $\mathbf{C} \cap M_n(\{-1, 0, 1\})$. In the former case, the diagonal entries must all be 1, and in the latter, they must be 0 or 1. (But, then, we need only check the principal submatrix based upon the diagonal entries that are equal to 1.) But, in general, the case in which all are 1 has seen the most interest. These matrices have a nice characterization [HP73]. A lemma of independent interest is useful.

Lemma 6.1.3 Let $A^T = A \in M_n(\mathbb{R})$ have all diagonal entries equal to 1. If every principal submatrix of A, in which all off-diagonal entries are strictly less than 1, is copositive, then A is copositive.

Proof If every off-diagonal entry of A is strictly less than 1, then the hypothesis includes the desired conclusion ($A \in \mathbf{C}$), and there is nothing to do. If not, suppose that $a_{ij} \geq 1$ and that $x \in \mathbb{P}_n$. Then, we show that there is a $y = [y_k] \in \Delta_n$, with $y_i y_j = 0$, such that $y^T Ay \leq x^T Ax$. Successive application of this observation yields that any value of the quadratic form of A on Δ_n by sums of values of the quadratic form (via nonnegative vectors) on principal submatrices in which all off-diagonal entries are < 1. Because these are all copositive, it will follow that A is copositive.

Suppose that $x \in \Delta_n, x_i x_j > 0$ and $a_{ij} \geq 1$. Without loss of generality, we can assume $i < j$. We construct $y \in \Delta_n$ such that $y_i y_j = 0$ and $y^T Ay \leq x^T Ax$. Let $t = x_i + x_j$ and define

$$x(u) = [x_1 \ \dots \ x_{i-1} \ \ u \ \ x_{i+1} \ \dots \ x_{j-1} \ \ t-u \ \ x_{j+1} \ \dots \ x_n]^T.$$

Now, $f(u) = x(u)^T Ax(u)$ is a concave function of u on $[0, t]$ (linear, if $a_{ij} = 1$). So the minimum of f, $0 \leq u \leq t$, occurs at an endpoint of $[0, t]$. Let y be the corresponding vector, and the necessary observation is proven. □

Theorem 6.1.4 Suppose that $A \in M_n(\{-1, 1\})$ with all diagonal entries 1. Then $A \in \mathbf{C}$, unless there is a 3-by-3 principal submatrix of A that is

$$\begin{bmatrix} 1 & -1 & -1 \\ -1 & 1 & -1 \\ -1 & -1 & 1 \end{bmatrix}.$$

Proof By the Lemma 6.1.3, we need only check principal submatrices, all of whose off-diagonal entries are -1. If there is one of size 3-by-3 or more, there is one of size 3-by-3. But the displayed matrix is not in $\mathbf{C}$, by evaluating the quadratic form at e; and, then, $A \notin \mathbf{C}$. On the other hand, if there is no such principal submatrix of size 3-by-3 or more, then $A \in \mathbf{C}$ by the Lemma 6.1.3. □

Theorem 6.1.5 Suppose that $A \in M_n(\{-1, 0, 1\})$ with all diagonal entries equal to 1. Then $A \in \mathbf{C}$, unless there is a 3-by-3 principal submatrix of A that is either

$$\begin{bmatrix} 1 & -1 & -1 \\ -1 & 1 & -1 \\ -1 & -1 & 1 \end{bmatrix}$$

or permutation similar to

$$\begin{bmatrix} 1 & -1 & -1 \\ -1 & 1 & 0 \\ -1 & 0 & 1 \end{bmatrix}.$$

Proof Let B be the first displayed matrix in the hypothesis and C be the second. Suppose A is copositive. Because neither B nor C is copositive, neither can be a principal submatrix of A. On the other hand, if A is not copositive, it follows from the lemma that A contains a principal submatrix M, which is not copositive and has all off-diagonal entries 0 or -1. If M contains a row with at least two -1s, then M contains a principal submatrix that is permutation similar to B or C. If each row of M contains at most one -1, then M is clearly the direct sum of matrices of the form

$$\begin{bmatrix} 1 & -1 \\ -1 & 1 \end{bmatrix} \quad \text{and} \quad \begin{bmatrix} 1 \end{bmatrix}.$$

So M is positive semidefinite, hence, copositive. □

In the ensuing paragraphs, we summarize the basic theory of cones and their connections with copositive matrices.

The topological interior of **C** is **SC**. Equivalently, the topological closure of **SC** is **C**. **C** is a closed convex cone in the class of symmetric matrices. **C** has a nonempty interior and is pointed, i.e., $\mathbf{C} \cap (-\mathbf{C}) = \{0\}$. **C** is not polyhedral, i.e., it is not the case that **C** is the intersection of a finite number of half-spaces.

A copositive matrix Q is an **extreme (copositive) matrix** if and only if whenever $Q = Q_1 + Q_2$, in which Q_1 *and* Q_2 are copositive, then $Q_1 = tQ$ and $Q_2 = (1-t)Q$ for some $t \in [0, 1]$. Other work in the 1960s focused on extreme copositive quadratic forms. The **polar (dual)** K^* of a cone $K \subseteq \mathbb{R}^n$ is the set of all $x \in \mathbb{R}^n$ such that $y^T x \geq 0$ for all $y \in K$. K is said to be **self-polar** if $K = K^*$. A matrix is **copositive with respect to a cone** K if $x^T Ax \geq 0$ for all $x \in K$.

6.2 Basic Properties of Copositive Matrices in $M_n(\mathbb{R})$

The copositive matrices form a bigger class than the **PSD** matrices, but they are a natural generalization that is analogous in many ways. We record some basic properties here.

Lemma 6.2.1 (Inheritance) Let $A \in M_n(\mathbb{R})$ and $\alpha \subseteq \{1, 2, \ldots, n\}$. If $A \in \mathbf{C}$ (**SC**), then the principal submatrix $A[\alpha] \in \mathbf{C}$ (**SC**).

Proof Let $A \in \mathbf{C}$ of order n and suppose there exists a vector $x \geq 0$, $x \neq 0$ such that $x^T A[\alpha]x \ngeq 0$. Then, for the n-by-1 vector $y = \left[\begin{smallmatrix} x \\ 0 \end{smallmatrix}\right]$,

$$\begin{aligned} y^T Ay &= \begin{bmatrix} x^T & 0 \end{bmatrix} \begin{bmatrix} A[\alpha] & A[\alpha, \alpha^c] \\ A[\alpha^c, \alpha] & A[\alpha^c] \end{bmatrix} \begin{bmatrix} x \\ 0 \end{bmatrix} \\ &= x^T A[\alpha]x \ngeq 0, \end{aligned}$$

a contradiction. The strictly copositive case is similar. □

Corollary 6.2.2 If $A \in \mathbf{C}$ (**SC**), then the diagonal entries of A are nonnegative (positive), and so, $\mathrm{Tr}(A) \geq 0$ ($\mathrm{Tr}(A) > 0$).

Because $x^T(A + B)x = x^T Ax + x^T Bx$ and $x^T(tA)x = tx^T Ax$ when $t \in \mathbb{R}$, we have

Lemma 6.2.3 If $A, B \in M_n(\mathbb{R})$ and $A, B \in \mathbf{C}$ (**SC**), then $A + B \in \mathbf{C}$ (**SC**) and $tA \in \mathbf{C}$ (**SC**) for $t > 0$. Thus, **C** and **SC** both form a cone in $M_n(\mathbb{R})$. Of course, $\mathbf{SC} \subseteq \mathbf{C}$.

Example 6.2.4 Notice that

$$A = \begin{bmatrix} 0 & 1 \\ 1 & 1 \end{bmatrix}$$

lies in **C**, but not **SC**. For $x = \left[\begin{smallmatrix} 1 \\ 0 \end{smallmatrix}\right]$, $x^T Ax = 0$. Also, $A \notin \mathbf{C}^+$, as $Ax \neq 0$.

Example 6.2.5 Notice that

$$A = \begin{bmatrix} 1 & -1 \\ -1 & 1 \end{bmatrix}$$

lies in **C**, as it is **PSD**, but not **SC** as $e^T Ae = 0$, where $e = [1 \quad 1]^T$. However, $A \in \mathbf{C}^+$, as $x^T Ax = 0$ implies that x is a multiple of e, and $Ae = 0$.

Because the vector $e > 0$, and $e^T Ae$ is just the sum of the entries of A, we have

Lemma 6.2.6 If $A \in M_n(\mathbb{R})$ and $A \in \mathbf{C}$ (**SC**), then the sum of the entries of A is nonnegative (positive).

Recall that a transformation in which $A \in M_n(\mathbb{R}) \to C^T AC$ for an invertible $C \in M_n(\mathbb{R})$ is called a **congruence of** A. If A is **PSD**, any congruence of A is again **PSD**. However, this is not so for **C**.

Example 6.2.7 If

$$A = \begin{bmatrix} 1 & 2 \\ 2 & 1 \end{bmatrix},$$

then A is strictly copositive, due to positivity. However, if

$$D = \begin{bmatrix} 1 & 0 \\ 0 & -1 \end{bmatrix},$$

then

$$D^T AD = \begin{bmatrix} 1 & -2 \\ -2 & 1 \end{bmatrix},$$

which is not copositive.

So, to maintain copositivity, restrictions must be made on congruence, especially if we wish to have an automorphism. First, because we wish the matrix carrying the congruence to map nonnegative vectors to nonnegative vectors, we have

Lemma 6.2.8 If $A \in M_n(\mathbb{R}) \cap \mathbf{C}$ and $C \in M_n(\mathbb{R})$, $C \geq 0$, then $C^T AC \in \mathbf{C}$.

Proof $x^T\left(C^TAC\right)x = (Cx)^TA(Cx) \geq 0$ because $Cx \geq 0$. So $C^TAC \in \mathbf{C}$. □

Recall the definition from Chapter 3 that $C \in M_{r,m}(\mathbb{R})$ is called row positive if $C \geq 0$ and C has at least one positive entry in each row. Similarly, for **SC**, we have

Lemma 6.2.9 If $A \in M_n(\mathbb{R}) \cap \mathbf{SC}$ and $C \in M_n(\mathbb{R})$ is row positive, then $C^TAC \in \mathbf{SC}$.

Proof Suppose $0 \neq x \geq 0$, and consider $x^TC^TACx = y^TAy$, for $y = Cx$. Because $Cx \geq 0$, $Cx \neq 0$ because C is row positive, $y^TAy > 0$, and C^TAC is **SC**. □

Remark 6.2.10 For every $x \in \mathbb{R}^n$ with some negative entries, there is an $A \in \mathbf{SC}$ such that $x^TAx < 0$. To see this, it suffices to consider $n = 2$ and A with small positive diagonal entries and a large positive off-diagonal entry.

Recall that a **monomial matrix** is a square matrix with exactly one positive entry in each row and column. Now, as nonnegative monomial matrices are those that map the nonnegative orthant onto itself, we have

Theorem 6.2.11 If $A, C \in M_n(\mathbb{R})$ and C is monomial, then $C^TAC \in \mathbf{C}\,(\mathbf{SC})$ if and only if $A \in \mathbf{C}\,(\mathbf{SC})$. Moreover, no broader class of congruences is an automorphism (of either class).

Proof By Lemma 6.2.9, $A \in \mathbf{C}\,(\mathbf{SC})$ implies $C^TAC \in \mathbf{C}\,(\mathbf{SC})$, as monomial matrices are row positive. Because the inverse of a monomial matrix is monomial, the converse follows, as well. Because the monomial matrices are the largest class of matrices F such that both F and F^{-1} map the nonnegative orthant to itself, Remark 6.2.10 ensures that no larger class of congruences works. □

Two useful special cases are the following.

Corollary 6.2.12 If $A \in M_n(\mathbb{R})$ and P is a permutation matrix, then $P^TAP \in \mathbf{C}\,(\mathbf{SC})$ if and only if $A \in \mathbf{C}\,(\mathbf{SC})$.

Corollary 6.2.13 If $A \in M_n(\mathbb{R})$ and $D \in D_n^+$, then $D^TAD \in \mathbf{C}\,(\mathbf{SC})$ if and only if $A \in \mathbf{C}\,(\mathbf{SC})$.

Because the diagonal entries must be positive (nonnegative) in order that a matrix be **SC** (**C**), the positive diagonal congruence automorphism may be used to normalize the diagonal entries of a matrix to be classified as **SC** (resp., **C**) to all be 1 (resp., 0 or 1). In the case of copositivity, it is useful to note that if a diagonal entry is 0, then the off-diagonal entries in that row and column must be nonnegative. This is an analog of a familiar fact about 0 diagonal entries of **PSD** matrices: the corresponding rows and columns must be 0.

Theorem 6.2.14 If $A = [a_{ij}] \in M_n(\mathbb{R}) \cap \mathbf{C}$ and $a_{ii} = 0$, then $a_{ij}, a_{ji} \geq 0, j = 1, 2, \ldots, n$.

Proof If $a_{ii} = 0$ and $a_{ij} < 0$, then the principal submatrix

$$A[\{i,j\}] = \begin{bmatrix} 0 & -s \\ -s & t \end{bmatrix}$$

with $s > 0$ and $t \geq 0$. By Lemma 6.2.1, this 2-by-2 matrix must be in **C**. But, its quadratic form on $\left[\begin{smallmatrix} x \\ y \end{smallmatrix}\right] \in \mathbb{R}^2$ is $ty^2 - 2sxy$. Choosing $y > 0$ and $x > \frac{ty}{2s}$ makes the quadratic form negative, a contradiction that completes the proof. □

Because of the inheritance requirement in Lemma 6.2.1, this means that $A \in M_n(\mathbb{R})$ is in **C** if and only if off-diagonal entries are nonnegative in rows with 0 diagonal entries and the principal submatrix induced by its positive diagonal entries is in **C**. Of course, if all diagonal entries are 0, the matrix must be nonnegative and, thus, copositive. This means that to check membership in **C** or **SC**, we need only consider matrices with all diagonal entries 1.

We have seen that each of the three classes **C**, **SC**, and $\mathbf{C}^+$ is closed under addition, positive scalar multiplication (thus, each is a cone), principal submatrix extraction (inheritance), and monomial congruence (thus, closed under permutation similarity). Copositive matrices are not closed under inversion or Schur complementation. Conditions for the inverse to be copositive can be found in [HS10, lemma 7.7]. A sufficient condition for the Schur complement to be copositive can be found in [JR05, theorem 4].

Example 6.2.15 If

$$A = \begin{bmatrix} 1 & 2 & 0 \\ 2 & 1 & 2 \\ 0 & 2 & 1 \end{bmatrix},$$

then $A \in \mathbf{C}$, the $(2,2)$ entry of $A^{-1} = -1/7$ and $A/A[\{1,2\}] = -5/3$. So $A^{-1} \notin \mathbf{C}$ and $A/A[\{1,2\}] \notin \mathbf{C}$.

6.3 Characterizations of C, SC, and C$^+$ Matrices

In the fundamental copositive paper [CHL70b], **C**(**SC**, $\mathbf{C}^+$) matrices are characterized using inductive criteria for copositivity. In [Had83], a more general criterion provides simpler coordinate-free proofs of the results of [CHL70b]. To begin, we simply state the main results of [CHL70b] (because the proofs are a bit tedious).

Definition 6.3.1 [Had83] The symmetric matrix $A \in M_n(\mathbb{R})$ is called **(strictly) copositive of order** m, $1 \le m \le n$, if and only if every principal submatrix of order m is (strictly) copositive. There are corresponding definitions with respect to **PD** and **PSD**.

Theorem 6.3.2 [CHL70b] Let $M \in M_n(\mathbb{R})$. Then $M \in \mathbf{C}^+$ if and only if for some permutation matrix P,

$$P^T MP = \begin{bmatrix} A & B \\ B^T & D \end{bmatrix}$$

in which

(i) $A \in M_r(\mathbb{R})$ is **PSD**, $0 \le r \le n$;
(ii) $B = A\hat{B}$ for some $\hat{B}$;
(iii) $D - \hat{B}^T A\hat{B} \in \mathbf{SC}$ (hence $D \in \mathbf{SC}$).

Theorem 6.3.3 [CHL70b] Let the symmetric matrix $M \in M_n(\mathbb{R})$ be copositive of order $n-1$. Then $M \notin \mathbf{C}$ if and only if

(i) $\operatorname{adj} M \ge 0$, and
(ii) $\det M < 0$

(in which case, M is **PSD** of order $n-1, M^{-1} \le 0$, and M has one negative eigenvalue that is of minimum magnitude and has a positive eigenvector).

Example 6.3.4 Consider the symmetric matrix

$$M = \begin{bmatrix} 1 & -1 & -1 \\ -1 & 1 & -1 \\ -1 & -1 & 1 \end{bmatrix},$$

which is **C** (but not **SC**) of order 2. Then

$$\operatorname{adj} M = \begin{bmatrix} 0 & 2 & 2 \\ 2 & 0 & 2 \\ 2 & 2 & 0 \end{bmatrix} \ge 0, \quad \text{and}$$

$\det M = -4 < 0$ so that $M^{-1} \le 0$. From Theorem 6.3.3, it follows that $M \notin \mathbf{C}$. Further, $M \in$ **PSD** of order 2 and M has exactly one negative eigenvalue, namely -1, of minimum modulus and with corresponding positive eigenvector e, the all 1s vector.

Similarly, we have

Theorem 6.3.5 [CHL70b, theorem 5.3.5] Let the symmetric matrix $M \in M_n(\mathbb{R})$ be **SC** of order $n-1$. Then $M \notin$ **SC** if and only if

(i) $\operatorname{adj} M > 0$, and
(ii) $\det M \leq 0$

(in which case, M is **PD** of order $n-1$, so that $\operatorname{rank}(M) \geq n-1$).

Furthermore, if $M \notin$ **SC**, then either

(a) $M \notin$ **C**, which is true if and only if $\det M < 0$ (in which case M has exactly one negative eigenvalue of strictly minimum magnitude with a positive eigenvector), or
(b) $M \in$ **C**, which is true if and only if $\det M = 0$ (in which case $M \in$ **PSD**, $\operatorname{rank} M = n-1$, and the eigenvector associated with 0 is positive).

Example 6.3.6 Consider the symmetric matrix

$$M = \begin{bmatrix} 3 & -2 & -1 \\ -2 & 2 & 0 \\ -1 & 0 & 1 \end{bmatrix}.$$

M is **PD** (and hence, **SC**) of order 2, $\det M = 0$, and $\operatorname{adj} M = 2J$ (in which J is the all 1s matrix). So, by contraposition, $M \in$ **C** and the eigenvector associated with 0 is $v = e$.

The following definition will be of use.

Definition 6.3.7 [Had83] Let A be a symmetric matrix, and let B be a strictly copositive matrix of the same order. The pair A, B is called a **(strictly) codefinite pair** if and only if $Ax = \lambda Bx, x > 0$ implies $\lambda \geq 0$ ($\lambda > 0$).

Now we establish the results of [Had83] for **C** (**SC**) matrices.

Theorem 6.3.8 [Had83] Let $A \in M_n(\mathbb{R})$ be copositive of order $n-1$. The following statements are equivalent:

(i) A is strictly copositive.
(ii) There is a strictly copositive matrix B of order n such that the pair A, B is (strictly) codefinite.
(iii) For every strictly copositive matrix B of order n the pair A, B is (strictly) codefinite.

Proof

(i) $\Rightarrow$ (iii): Let B be **SC** of order n. Suppose $Ax = \lambda Bx, x > 0$. Then, $x^T Ax = \lambda x^T Bx$, and thus $\lambda \geq 0 (\lambda > 0)$.

(iii) $\Rightarrow$ (ii): This is obvious.

(ii) $\Rightarrow$ (i): Suppose the form $x^T Ax$ assumes negative (nonpositive) values for certain $x > 0, x \neq 0$. Because A is (strictly) copositive of order $n-1$, such x are necessarily > 0. Consider the form $x^T Ax$ on the set $D = \{x\colon x \geq 0, x^T Bx = 1\}$. It assumes its minimum at a point $\hat{x}$ of the relative interior of D. The minimum is negative (nonpositive). From the necessary condition for a minimum of $x^T Ax$ on D it follows that $A\hat{x} = \lambda B\hat{x}$ with $\lambda = \hat{x}^T A\hat{x}$. From (ii) $\lambda \geq 0$(or $\lambda > 0$), which leads to a contradiction. □

The next theorem is the main result of [CHL70a].

Theorem 6.3.9 [Had83] Let $A \in M_n(\mathbb{R})$ be copositive of order $n-1$. Then the following statements are equivalent:

(i) $A \notin \mathbf{C}$.
(ii) For every $b > 0$ there is an $x > 0$ such that $Ax = \lambda b,\ \lambda < 0$.
(iii) The matrix $-A^{-1}$ exists and is nonnegative.
(iv) $\det A < 0$ and $\operatorname{adj} A \geq 0$.

Proof For the matrix B in Theorem 6.3.8 one can choose the identity or matrices $B = bb^T$ of rank 1 in which the vector $b > 0$.

(i) $\Leftrightarrow$ (ii): This follows from Theorem 6.3.8, by contraposition.

(ii) $\Rightarrow$ (iii): Assume that (ii) holds and that there is a y such that $Ay = 0$. Because $Ax = 0$ has a solution for $b > 0$, it follows that $y^T b = 0$ for $b > 0$ so that $y = 0$. Hence $-A^{-1}$ exists. For every $b > 0$ the vector $x = A^{-1}b < 0$. Thus, $-A^{-1} \geq 0$, establishing (iii).

(iii) $\Rightarrow$ (ii): Assume that (iii) holds and that $b > 0$. Then let $\lambda = -1$ and $x = -A^{-1}b$ so that $Ax = A(-A^{-1}b) = -b = \lambda b, \lambda < 0$.

(iv) $\Rightarrow$ (iii): This is obvious.

(iii) $\rightarrow$ (iv): Assume that (iii) holds. Choose $B = I$ in Theorem 5.3.8. Then there is $x > 0$ such that $Ax = \lambda x, \lambda < 0$. Assume that A has a second negative eigenvalue, $Ay = \mu y, y^T x = 0, y \neq 0, \mu < 0$. Therefore, the vector y has positive and negative components because $x > 0, y^T x = 0$. Hence, there is $\alpha > 0$ such that $x + \alpha y \geq 0$ and $x + \alpha y$ has at least one zero component. Because A is copositive of order $n-1$,

$$0 \leq (x+\alpha y)^T A(x+\alpha y) = \lambda x^T x + \mu \alpha^2 y^T y < 0,$$

a contradiction. Thus, A has exactly one negative eigenvalue, which implies $\det A < 0$. (iv) then follows from (iii). □

For completeness, the corresponding result for strictly copositive matrices is included. The proof is similar to that of Theorem 6.3.9.

Theorem 6.3.10 [Had83] Let $A \in M_n(\mathbb{R})$ be strictly copositive of order $n-1$. Then the following statements are equivalent:

(i) $A \notin$ **SC**.
(ii) For every $b > 0$, there is an $x > 0$ such that $Ax = \lambda b,\ \lambda \leq 0$.
(iii) The matrix $-A^{-1}$ exists and is negative.
(iv) $\det A \leq 0$ and $\operatorname{adj} A > 0$.

For a symmetric matrix $A \in M_n(\mathbb{R})$, $Q(x) = x^T Ax$ denotes its **associated quadratic form**, $\|x\|$ denotes the Euclidean norm of x, and $\mathbb{R}^n_+$ denotes the set of all x in $\mathbb{R}^n$ such that $x \geq 0, x \neq 0$. In the following a spectral criterion for strict copositivity is established.

Theorem 6.3.11 [Kap00] Let $A \in M_n(\mathbb{R})$ be a symmetric matrix. Then A is strictly copositive if and only if every principal submatrix B of A has no eigenvector $v > 0$ associated with an eigenvalue $\lambda \leq 0$.

Proof Let the principal submatrices of A have the property stated in the theorem, and let $Q(x_0) \leq 0$ for some x_0 in $\mathbb{R}^n_+$. Some, but not all, components of x_0 may be 0. For proper numbering, we can assume that $x_0 = [a_1\ \ldots\ a_m\ \ 0\ \ldots\ 0]^T$, where $1 \leq m \leq n$, and $a_i > 0$ for $i = 1, \ldots, m$. The vector x_0 cannot be the unique one at which Q has a nonpositive value; among all such vectors, we can choose one at which m has its least value, and we assume that our x_0 has this property.

We consider the case $m > 1$. Let $y = [y_k]$ be an arbitrary vector of $\mathbb{R}^m$ and let

$$Q_0(y) = Q(y_1, \ldots, y_m, 0, \ldots, 0).$$

Let $y_o = [a_1\ \ldots\ a_m]^T$. We consider the function $Q_0(y)$ restricted to the set $E = \{y \in \mathbb{R}^m\ :\ \| y \| = 1, y \geq 0\}$. By multiplying y_0 by a positive scalar, we can ensure that y_0 is in E. The vector y_0 must then be in the relative interior of E because a vector in the relative boundary of E would have less than m nonzero components. Thus, the function $Q_0(y)$ on E has positive values on the relative boundary of E and has a nonpositive value at one point in the relative interior. Accordingly, the function must have a negative or zero absolute minimum at some vector v in the relative interior, so that $v > 0$. But the vector v would be an eigenvector of a principal submatrix of A with a negative or a zero associated eigenvalue. By hypothesis, there can be no such vector. Accordingly, $Q(x) > 0$ in $\mathbb{R}^n_+$.

For the case $m = 1$, the principal submatrix $[a_{11}]$ would have the eigenvector $v = (1)$ with a nonpositive eigenvalue, and the same conclusion follows.

Conversely, let $Q(x) > 0$ in $\mathbb{R}^n_+$, and let some principal submatrix B of A have an eigenvector $v > 0$ with a nonpositive associated eigenvalue. We can assume that B is obtained from A by deleting the rows and columns following the m-th and write

$$v = [a_1 \ \dots \ a_m]^T, \quad x_0 = [a_1 \ \dots \ a_m \ \ 0 \dots 0]^T.$$

Then $Q(x_0) = v^T B v \leq 0$, contrary to hypothesis. Therefore, no principal submatrix of A can have an eigenvector $v > 0$ with a negative or a zero associated eigenvalue. This establishes the theorem. □

Note that the matrix M of Example 6.3.4 has eigenvector

$$v_1 = \begin{bmatrix} 2 \\ 2+\sqrt{6} \\ 2+\sqrt{6} \end{bmatrix} > 0$$

with associated eigenvalue $\lambda_1 = 2 - \sqrt{6} \leq 0$. Hence, $M \notin$ **SC**.

For copositivity, an analogous proof yields

Theorem 6.3.12 [Kap00] Let $A \in M_n(\mathbb{R})$ be a symmetric matrix. Then A is copositive if and only if every principal submatrix B of A has no eigenvector $v > 0$ associated with an eigenvalue $\lambda < 0$.

For completeness, we state Motzkin's determinantal test for strict copositivity [CHL70b], as well as that of the corresponding test for copositivity due to Keller.

Theorem 6.3.13 (Motzkin) [CHL70b] A symmetric matrix is strictly copositive if and only if each principal submatrix for which the cofactors of the last row are all positive has a positive determinant. (This includes the positivity of the diagonal entries.)

Theorem 6.3.14 (Keller) [CHL70b] A symmetric matrix is copositive if and only if each principal submatrix for which the cofactors of the last row are all nonnegative has a nonnegative determinant.

For example, the matrix M in Example 6.3.4 has cofactors $2, 2, 3$ in the last row and $\det M = -4 < 0$. Hence, M is not copositive.

For $n \leq 3$, explicit entry-wise inequalities are given [Had83] that provide necessary and sufficient conditions for a matrix to lie in **C**(**SC**). There are such formulas for $n = 4$, but the inequalities are much more complicated [PF93].

6.4 History

Now we summarize a number of papers on copositivity.

In [Bau66], it was shown that for $n \geq 2$, if $Q(x) = x^T Qx$ in which $x = [x_1 \dots x_n]^\top \geq 0$ is an extreme copositive quadratic form, then for any i, $i = 1, \dots, n$, Q has a zero $u = [u_1 \dots u_n]^T \geq 0$ in which $u_i = 0$. It was also shown that extreme forms in n variables can be used to construct extreme forms in $m > n$ variables.

In [Bau67] a class of extreme copositive forms in five variables was introduced and shown to provide extreme copositive forms in any number of variables $m > n$.

In [Bas68] and [HP73], a number of results were established about relations with the class $\mathcal{H}$ of symmetric matrices in which every entry is ± 1. In [HP73], these results were extended to the class $\mathcal{E}$ of symmetric matrices in which every entry is ± 1 or 0, and each diagonal entry is 1. Specifically, those matrices in $\mathcal{E}$, which are, respectively, copositive, copositive-plus, or positive semidefinite, are characterized.

In [HN63] it was shown that the class **C**of copositive forms and the class $\mathcal{B}$ of completely positive forms (i.e., real forms that can be written as a sum of squares of nonnegative real forms) are convex cones that are dual to each other and that **PSD** and **sN** are self-dual. Also, the extreme copositive forms of **PSD**, **sN**, $\mathcal{B}$, and **PSD** + **sN** were determined, as well as some extreme forms of **C**. Horn's form is an example of an extreme copositive form that does not belong to **PSD** + **sN**. In [Bau66] necessary and sufficient conditions for a quadratic form $x^T Ax$ with $|a_{ij}| = 1$ to be an extreme copositive form are obtained.

Let S denote the set of symmetric matrices with 1s on the diagonal and off-diagonal entries 0 or ± 1. In [HN63] the matrices in S that lie in **C** (resp. $\mathbf{C}^+$, **PSD**) are characterized, as well as the copositive matrices in S that lie on extreme rays of **C**.

Let S_0 denote the set of symmetric matrices with 1s on the diagonal and with ± 1 off-diagonal. Matrices in S_0 that are (1) copositive, (2) extreme rays of the cone of copositive matrices, and (3) positive semidefinite are characterized in terms of certain graphs [HP73].

Copositivity has surfaced in a variety of areas of applied mathematics, which include game theory [Bas68, Gad58, Lan90], mathematical programming

[Ber81, CHL70a, CHL70b, Val86, Val88, Val89a, Val89b], the theory of inequalities [Dia62], block designs [HN63], optimization problems [Dan90], and control theory [HS10]. More recently there has been renewed interest in copositivity due to applications in modeling [CKL92, Had83] and linear complementarity problems [Ber81, MND01]. In particular, copositive matrices are of interest in solving the following linear complementarity problem as it arises in game theory [CHL70a]: Find a solution z to the system $w = q + Mz$, where M is an n-by-n copositive matrix, subject to $w \geq 0$, $z \geq 0$, $z^T w = 0$.

In [Val86], the basic theory of copositive matrices is reviewed and supplemented. Finite criteria for **C**, $\mathbf{C}^+$, and **SC** matrices are derived using principal pivoting and compared with existing determinantal tests for $\mathbf{C}^+$matrices. Last, $\mathbf{C}^+$matrices of order 3 are characterized.

In [Val89a], a real n-by-n matrix is defined to be **almost C** (**almost** $\mathbf{C}^+$, **almost SC**) if it is not $\mathbf{C}(\mathbf{C}^+, \mathbf{SC})$, but all its principal submatrices of order $n-1$ are. Using principal pivoting and quadratic programming, the above classes of matrices are studied with necessary and sufficient conditions given for almost $\mathbf{C}^+$matrices.

In [Val89b], criteria based on quadratic programming are given for **C**, $\mathbf{C}^+$, and **C** matrices and these are compared with existing criteria.

The copositivity of a symmetric matrix of order n is shown to be equivalent to the copositivity of two symmetric matrices of order $n-1$, provided that the matrix has a row whose off-diagonal entries are nonpositive [PF93]. Based on this result, criteria are derived for copositive matrices of orders 3 and 4.

As one might expect, one of the primary areas of interest has been in detection, i.e., given a general (square) matrix, determining whether it lies in the cone of copositive matrices. This is an NP-hard problem [Bom96]. Papers on detection include [Dan90, CHL70b, Kap01, HS10, Bas68, Bom00, Bom08, Bom87, Bom96]. Additional algorithms for detection of copositive matrices were developed in [XY11, Bom00, Bom96]. In [ACE95] simplices and barycentric coordinates are used in determining criteria for copositive matrices.

Special copositive matrices were introduced in [Gow89, MND01, MP98]. For the quadratic form $x^T Qx$ to be positive unless $x = 0$, subject to the linear inequality constraints $Ax \geq 0$, it is sufficient that there exists a copositive matrix C such that $Q - A^T CA$ is positive definite [MJ81]. The main result of [MJ81] shows this condition is also necessary.

Recent survey papers are [HS10] (on variational aspects of copositive matrices), [IS00] (for conditionally definite matrices), [Dur10] (for copositive programming), [BS-M03] (for completely copositive cones), and [Dic11] (for the geometry of the copositive cone and its dual cone – the cone of completely positive matrices). The interior of the completely positive cone is characterized in [Dic10], using the set of zeros in the nonnegative orthant of the form.

The (strictly) copositive matrix completion problem was solved in [HJR05].

6.5 Spectral Properties

In 1969, the earliest spectral result on copositive matrices was obtained [HH69]. Specifically, it was shown that a copositive matrix A has the Perron property, i.e., its spectral radius $\rho(A)$ is an eigenvalue. In 2005, among other things, it was shown that a copositive matrix must have a positive vector in the subspace spanned by the eigenvectors corresponding to the nonnegative eigenvalues [JR05]. Moreover, in 2008, necessary conditions were given that ensure that the Perron root of a copositive matrix has a positive eigenvector [Bom08].

Theorem 6.5.1 [HH69] Let A be a copositive matrix. Then its largest eigenvalue r satisfies $r \geq |\lambda|$, in which λ is any other eigenvalue of A, i.e., A has the Perron property.

Proof Let $\lambda < 0$ and $Ax = \lambda x$ in which $||x|| = 1$. Then write $x = y - z$ in which $y \geq 0$, and $y^T z = 0$, so $||y + z|| = 1$. Then

$$r \geq (y+z)^T A(y+z) = 2y^T Ay + 2z^T Az - (y-z)^T A(y-z) \geq -\lambda = |\lambda|.$$

□

6.6 Linear Preservers

Recall the opening paragraph of Section 3.6 for the notion and utility of studying the linear preservers of a matrix class. In this section, which is based on [FJZ19], we are interested in identifying linear transformations on symmetric matrices that preserve copositivity, either in the into or onto sense.

Let $\mathbf{S}_n = \mathbf{S}_n(\mathbb{R})$ be the $\frac{n(n+1)}{2}$-dimensional subspace of $M_n(\mathbb{R})$ consisting of symmetric matrices. In this section, we use a subscript to indicate the size of copositive matrices, if useful; that is, $\mathbf{C}_n$ and $\mathbf{SC}_n$ denote $\mathbf{C} \cap \mathbf{S}_n$ and $\mathbf{SC} \cap \mathbf{S}_n$, respectively.

In considering the action of linear transformations on copositive matrices, it suffices to consider linear transformations L: $\mathbf{S}_n \to \mathbf{S}_n$ (and it can be convenient because there are fewer variables to consider). However, it may also be convenient to consider L to be a linear map on $M_n(\mathbb{R})$. We shall do so interchangeably. We say that such a linear transformation **preserves copositivity** if $A \in \mathbf{C}$ implies $L(A) \in \mathbf{C}$, and similarly for strict copositivity. More precisely, such an L is an **into copositivity preserver**. If $L(\mathbf{C}) = \mathbf{C}$, we have an **onto copositivity preserver**. Our purpose here is to better understand both types of linear copositivity preservers. The into preservers of the PSD matrices are not fully understood, and they are recognized to be a difficult problem. The onto linear preservers of PSD are straight forwardly known to be the congruences by a fixed invertible matrix [Schn65].

Certain natural kinds of linear transformations are more amenable to preserver analysis. We say that L is a **linear transformation of standard form** on $M_n(\mathbb{R})$ if there are fixed matrices $R, S \in M_n(\mathbb{R})$ such that $L(A) = RAS$ (or $L(A) = RA^T S$). Such a linear transformation is invertible if and only if R and S are invertible matrices. More generally, L is a linear transformation on $M_n(\mathbb{R})$ if $L(A) = [l_{ij}(A)]$, in which l_{ij} is a linear functional in the entries of A. It is known that an invertible linear transformation that preserves rank is of standard form [Hndbk06, Pier92], and there are useful variations upon this sufficient condition.

Both the "onto" and especially the "into" copositivity linear preserver problems appear subtle. For example, in [Pok] a conjecture of N. Johnston is relayed: Any (into) copositivity preserver is of the form

$$X \longrightarrow \sum_i A_i^T X A_i,$$

in which $A_i \geq 0$. Because any such map is (clearly) a copositivity preserver, this is a natural (though optimistic) conjecture. However, it is false even for 2-by-2 matrices. Suppose that a linear map on $\mathbf{S}_2$ is given by

$$\begin{bmatrix} a & c \\ c & b \end{bmatrix} \to \begin{bmatrix} a & a+b+2c \\ a+b+2c & b \end{bmatrix}.$$

If the argument is copositive, then $a + b + 2c \geq 0$ $\big($because it is the value of the quadratic form of the argument at $\left[\begin{smallmatrix} 1 \\ 1 \end{smallmatrix}\right]\big)$, so that the image is nonnegative and, thus, copositive. But, as the conjectured form is a PSD preserver (by virtue of being a sum of congruences), the fact that $\left[\begin{smallmatrix} 10 & -1 \\ -1 & 10 \end{smallmatrix}\right]$ is PSD, while its image $\left[\begin{smallmatrix} 10 & 18 \\ 18 & 10 \end{smallmatrix}\right]$ is not, shows that our linear map is not of the conjectured form, though a copositivity preserver.

6.6.1 Linear Copositivity Preservers of Standard Form

Because our arguments are always in $\mathbf{S}_n$, a linear transformation of standard form is of the form

$$L(A) = RAR^T, \quad \text{or } L(A) = RA^T R^T,$$

where $R \in M_n(\mathbb{R})$. Here we characterize both the into and onto preservers of copositivity of standard form. Such a transformation is invertible on $\mathbf{S}_n$ if and only if R is invertible. First, a useful lemma about copositive matrices. We say that $v \in \mathbb{R}^n$, $n \geq 2$, is a **vector of mixed sign** if v has both positive and negative entries. Nothing is assumed about 0 entries.

Lemma 6.6.1 For each vector $v \in \mathbb{R}^n$ of mixed sign, there is a matrix $A \in \mathbf{C}_n$ such that $v^T Av < 0$.

Proof By the permutation similarity invariance of $\mathbf{C}_n$, we may assume that $v = [v_1, v_2, \ldots, v_n]^T$ with $v_1 v_2 < 0$. Then, let $A \in \mathbf{C}_n$ be

$$\left[\begin{array}{cc|c} 0 & 1 & \\ 1 & 0 & 0 \\ \hline & 0 & 0 \end{array}\right],$$

so that $v^T Av = 2v_1 v_2 < 0$, as claimed. □

Theorem 6.6.2 Suppose that L is an into linear preserver of $\mathbf{C}_n$ of standard form. Then,

$$L(A) = S^T AS,$$

with $S \in M_n(\mathbb{R})$ and $S \geq 0$.

Proof Because L preserves symmetry and is of standard form, $L(A) = S^T AS$ with $S \in M_n(\mathbb{R})$. Suppose that there is an $x \geq 0$, $x \in \mathbb{R}^n$, such that Sx is of mixed sign. Then by Lemma 6.6.1, the argument A may be chosen so that $0 > (Sx)^T A(Sx) = x^T S^T ASx$ and $S^T AS = L(A)$ is not in $\mathbf{C}_n$. Hence, L is not a copositivity preserver. Thus, for any $x \geq 0$, Sx must be weakly uniformly signed. This means that $S \geq 0$ or ≤ 0. Because S appears twice, we may take it to be the former. □

Because $\mathbf{C}_n$ contains a basis of $\mathbf{S}_n$, a linear map on $\mathbf{S}_n$ that is an onto copositivity preserver must be an invertible linear map, and the inverse map must also be an onto preserver. If the map is of standard form, the inverse map just corresponds to inverting the S and S^T (which, of course, must be invertible). Thus, we have that $S^{-1} \geq 0$, as well as $S \geq 0$ (or $S^{-1} \leq 0$ and $S \leq 0$). It is well known that this happens if and only if S is a monomial matrix, the product of a permutation matrix and a positive diagonal matrix. Taken together, this gives the following characterization of linear transformations of standard form that map $\mathbf{C}_n$ onto itself.

Theorem 6.6.3 Suppose that L is an onto linear preserver of $\mathbf{C}_n$ of standard form. Then,

$$L(A) = S^T AS,$$

in which $S \in M_n(\mathbb{R})$ and S is monomial.

Because $\mathbf{C}_n$ forms a cone, we note that (i) any sum of into $\mathbf{C}_n$ preservers is again an into $\mathbf{C}_n$ preserver, though a sum of 1s of standard form may no longer be of standard form, and (ii) the sum of onto $\mathbf{C}_n$ preservers need no longer be onto. Also, it follows from the proven forms that both into and onto copositivity preservers of standard form are also (into and onto, respectively) PSD preservers.

6.6.2 Hadamard Multiplier $\mathbf{C}_n$ Preservers

Recall that the Hadamard, or entry-wise, product of two matrices $A = [a_{ij}]$ and $B = [b_{ij}]$ of the same size is defined and denoted by $A \circ B = (a_{ij}b_{ij})$. If we consider a fixed n-by-n matrix H, then a natural type of linear transformation on n-by-n matrices A is given by

$$L(A) = H \circ A. \tag{6.6.1}$$

We may also ask for which H are such transformations (into) $\mathbf{C}_n$ preservers.

Also recall that B is completely positive (**CP**) if $B = FF^T$ with F n-by-k and $F \geq 0$. Thus, for $F = [f_1, f_2, \ldots, f_k]$ partitioned by columns,

$$B = f_1 f_1^T + f_2 f_2^T + \cdots + f_k f_k^T.$$

Then, the **CP** matrices, which also form a cone, are special **PSD** matrices. It is known [HN63] that the n-by-n **CP** matrices are the cone theoretic dual of $\mathbf{C}_n$, as $Tr\left(B^T A\right) = \sum_{i=1}^k f_i^T A f_i$, which is nonnegative if $A \in \mathbf{C}_n$.

Now, consider a linear transformation of the form (6.6.1) with H a **CP** matrix of the form $H = \sum_{i=1}^k h_i h_i^T$, $h_i \in \mathbb{R}^n$, $h_i \geq 0$, $i = 1, \ldots, k$. If $A \in \mathbf{C}_n$, then $H \circ A = \sum_{i=1}^k h_i h_i^T \circ A$ and $x^T (H \circ A) x = \sum_{i=1}^k (x \circ h_i)^T A (x \circ h_i)$, which is nonnegative for $x \geq 0$.

Theorem 6.6.4 A linear transformation of the form (6.6.1) is an into copositive preserver if and only if H is **CP**.

Proof Sufficiency follows from the calculation above. The quadratic form of $H \circ A$ on a nonnegative vector is a sum of quadratic forms of A on nonnegative vectors. On the other hand, if H is not **CP** because of the known duality, $e^T (H \circ A) e = TrH^T A < 0$ for some $A \in \mathbf{C}_n$, and $H \circ A \notin \mathbf{C}_n$. □

If H is **CP**, $H = \sum_{i=1}^{k} h_i h_i^T$, $h_i \geq 0$, let $D_i = \mathrm{diag}(h_i)$. Then $H \circ A = \sum_{i=1}^{k} D_i^T A D_i$, with $D_i \geq 0$, so that a linear transformation of the form (6.6.1) is also a sum of into transformations of standard form. With the exception of H being a rank one, positive, symmetric matrix, such a transformation will not be onto.

6.6.3 General Linear Maps on $\mathbf{S}_n$

Let $L(A) = [l_{ij}(A)]$ in which each entry l_{ij} is a linear functional in the entries of A. Symmetry requires that the functionals l_{ij} and l_{ji} be the same. It is possible to design such maps that are copositivity preservers (and in a similar way, **PSD** preservers).

Let $z_{ij}^{(k)} \in \mathbb{R}^n$, $z_{ij}^{(k)} \geq 0$ and $z_{ij}^{(k)} = z_{ji}^{(k)}$. If $A \in \mathbf{C}_n$, then $z_{ij}^{(k)^T} A z_{ij}^{(k)} \geq 0$. Define

$$l_{ij}(A) = \sum_k z_{ij}^{(k)^T} A z_{ij}^{(k)}$$

and

$$L(A) = [l_{ij}(A)].$$

Then, for $A \in \mathbf{C}_n$, $L(A) \geq 0$ and $L(A)^T = L(A)$, so that $L(A) \in \mathbf{C}_n$ and L is an into $\mathbf{C}_n$ preserver. **PSD** (into) preservers may be designed in a similar way. For example, let $L(A) = \mathrm{diag}(l_1(A), \ldots, l_n(A))$ with $l_i(A) = z_i^* A z_i$, $z_i \in \mathbb{C}^n$, so that if A is **PSD**, $L(A)$ is a nonnegative diagonal matrix and, thus, **PSD**.

We note that, with this machinery, it is possible to design into, not onto, but invertible $\mathbf{C}_n$ preservers. Here is an example. For

$$A = \begin{bmatrix} a & b \\ b & c \end{bmatrix},$$

let

$$l_{11}(A) = \begin{bmatrix} 1 & 1 \end{bmatrix} A \begin{bmatrix} 1 \\ 1 \end{bmatrix} + \begin{bmatrix} 1 & 0 \end{bmatrix} A \begin{bmatrix} 1 \\ 0 \end{bmatrix},$$

$$l_{12}(A) = l_{21}(A) = \begin{bmatrix} 1 & 1 \end{bmatrix} A \begin{bmatrix} 1 \\ 1 \end{bmatrix},$$

and

$$l_{22}(A) = \begin{bmatrix} 1 & 1 \end{bmatrix} A \begin{bmatrix} 1 \\ 1 \end{bmatrix} + \begin{bmatrix} 0 & 1 \end{bmatrix} A \begin{bmatrix} 0 \\ 1 \end{bmatrix}.$$

Then,

$$L\left(\begin{bmatrix} a & b \\ b & c \end{bmatrix}\right) = \begin{bmatrix} 2a+2b+c & a+2b+c \\ a+2b+c & a+2b+2c \end{bmatrix}.$$

and, for $A \in \mathbf{C}_2$, $L(A) \geq 0$ and $L(A)$ is **PSD**. So L is a $\mathbf{C}_2$ preserver. Because $l_{12} \geq 0$, it is only into. However, L is invertible, as

$$L^{-1}\begin{bmatrix} x & z \\ z & y \end{bmatrix} = \begin{bmatrix} x-z & \frac{3z-x-y}{2} \\ \frac{3z-x-y}{2} & y-z \end{bmatrix}.$$

We note that more elaborate maps may be designed, including the possibility of negative off-diagonal entries.

6.6.4 PSD Preservers That Are Not Copositivity Preservers

Of course, a **PSD** preserver need not be a copositivity preserver, and we note here a famous example of a **PSD** preserver that is not a copositivity preserver.

The Choi map [Choi75] is a linear transformation from $M_3(\mathbb{R})$ to $M_3(\mathbb{R})$ that preserves **PSD**, but is not of any typical type. It is defined by

$$L\left(\begin{bmatrix} a_{11} & a_{12} & a_{13} \\ a_{21} & a_{22} & a_{23} \\ a_{31} & a_{32} & a_{33} \end{bmatrix}\right) = \begin{bmatrix} 2a_{11}+2a_{22} & -a_{12} & -a_{13} \\ -a_{21} & 2a_{22}+2a_{33} & -a_{23} \\ -a_{31} & -a_{32} & 2a_{33}+2a_{11} \end{bmatrix}.$$

It is known, and easily checked, that any 3-by-3 (symmetric) **PSD** matrix is transformed to another **PSD** matrix. Of course, L is not a fixed congruence, nor onto. However, note that a copositive matrix with 0 diagonal and positive off diagonal entries is transformed to a matrix that is not copositive.

In general, copositivity preservers need not be **PSD** preservers, and **PSD** preservers need not be copositivity preservers. However, we conjecture that onto copositivity preservers are also onto **PSD** preservers.

6.6.5 Onto Linear Copositivity Preservers

We first make a fundamental observation about onto preservers of $\mathbf{C}_n$.

Theorem 6.6.5 Let L: $\mathbf{S}_n \to \mathbf{S}_n$ be a linear transformation that maps $\mathbf{C}_n$ onto $\mathbf{C}_n$. Then L is invertible, and L^{-1} maps $\mathbf{C}_n$ onto $\mathbf{C}_n$.

Proof Because the copositive matrices include the standard basis for the symmetric matrices, their span is all of $\mathbf{S}_n$, which means that the map is invertible. Because $L(\mathbf{C}_n) = \mathbf{C}_n$, application of L^{-1} to both sides of the equality yields the desired statement. □

Now we may see that linear onto preservers of $\mathbf{C}_n$ also preserve several related sets. For $B \in \mathbf{C}_n$, let

$$\mathbf{C}(B) = \{A \in \mathbf{C}_n : \exists\alpha > 0 : B - \alpha A \in \mathbf{C}_n\}$$

and

$$\mathcal{R} = \{B \in \mathbf{C}_n : \mathbf{C}(B) = \mathbf{C}_n\}.$$

Corollary 6.6.6 Let $L : \mathbf{S}_n \to \mathbf{S}_n$ be a linear transformation that maps $\mathbf{C}_n$ onto $\mathbf{C}_n$. Then L also preserves (in the onto sense)

(a) the boundary copositive matrices;
(b) the interior of $\mathbf{C}_n$;
(c) $\mathbf{SC}_n$; and
(d) $\mathcal{R}$.

Proof Items (a) and (b) follow because the copositive matrices are the closure of the strictly copositive matrices, and an invertible linear transformation is continuous. Item (c) follows because $\mathbf{SC}_n$ is the interior of $\mathbf{C}_n$. Now we show (d). Let $B \in \mathcal{R}$ and $A \in \mathbf{C}$. Then $B - \alpha A \in \mathbf{C}$ for some $\alpha > 0$. So, $L(B - \alpha A) \in \mathbf{C}$, and $L(B) - \alpha L(A) \in \mathbf{C}$. Because L is onto, $L(A)$ runs over $\mathbf{C}$ when A runs over $\mathbf{C}$. So $L(B) \in \mathcal{R}$, and $L(\mathcal{R}) \subseteq \mathcal{R}$. Similarly, $L^{-1}(\mathcal{R}) \subseteq \mathcal{R}$. Thus, $\mathcal{R} \subseteq L(\mathcal{R}) \subseteq \mathcal{R}$ implying that $L(\mathcal{R}) = \mathcal{R}$. □

Because both a fixed permutation similarity (equivalently, congruence) or a fixed positive diagonal congruence is an onto copositivity preserver, we have that monomial congruence is an onto copositivity preserver.

Conjecture. The onto copositivity preservers are exactly the fixed monomial congruences.

In a moment, we prove this conjecture in the 2-by-2 case. However, a proof, without additional assumptions, appears subtle. We have already shown that the conjecture holds if

(1) the map is of standard form (Section 6.6.1). However, each of the following alternative additional assumptions is sufficient:

(2) the map is an (onto) **PSD** preserver;

(3) the map is rank nonincreasing (in which case it is of standard form [Loe89, MM59]);

(4) each of the component maps l_{ij} is a function of only one entry of A;

(5) each of the component maps l_{ij} has nonnegative coefficients.

We now study the onto 2-by-2 copositivity preservers. A symmetric matrix $A \in \mathbf{S}_2$ is copositive if and only if

$$A = \begin{bmatrix} a & b \\ b & c \end{bmatrix}$$

with $a \geq 0$, $c \geq 0$, and $b \geq -\sqrt{ac}$. The matrix A is strictly copositive if all the inequalities are strict.

From now on, we assume that $L\colon \mathbf{S}_2 \to \mathbf{S}_2$ is a linear transformation that maps $\mathbf{C}_2$ onto $\mathbf{C}_2$ and we use the following notation

$$L\left(\begin{bmatrix} 1 & 0 \\ 0 & 0 \end{bmatrix}\right) = \begin{bmatrix} \alpha_{11} & \alpha_{12} \\ \alpha_{12} & \alpha_{22} \end{bmatrix} =: \Pi_\alpha$$

$$L\left(\begin{bmatrix} 0 & 0 \\ 0 & 1 \end{bmatrix}\right) = \begin{bmatrix} \gamma_{11} & \gamma_{12} \\ \gamma_{12} & \gamma_{22} \end{bmatrix} =: \Pi_\gamma$$

$$L\left(\begin{bmatrix} 0 & 1 \\ 1 & 0 \end{bmatrix}\right) = \begin{bmatrix} \beta_{11} & \beta_{12} \\ \beta_{12} & \beta_{22} \end{bmatrix} =: \Pi_\beta,$$

so that

$$L\left(\begin{bmatrix} a & b \\ b & c \end{bmatrix}\right) = a\Pi_\alpha + b\Pi_\beta + c\Pi_\gamma.$$

Lemma 6.6.7 We have $\alpha_{11}\alpha_{22} = 0$ and $\gamma_{11}\gamma_{22} = 0$.

Proof We show the first claim. The proof of the second claim is analogous. According to Corollary 6.6.6, either $\alpha_{11}\alpha_{22} = 0$ or

$$\Pi_\alpha = \begin{bmatrix} \alpha_{11} & -\sqrt{\alpha_{11}\alpha_{22}} \\ -\sqrt{\alpha_{11}\alpha_{22}} & \alpha_{22} \end{bmatrix},$$

with $\alpha_{11} > 0$ and $\alpha_{22} > 0$. In order to get a contradiction, suppose that Π_α has this latter form.

- Suppose that $\beta_{11}\beta_{22} \neq 0$. Then, by Corollary 6.6.6,

$$\Pi_\beta = \begin{bmatrix} \beta_{11} & -\sqrt{\beta_{11}\beta_{22}} \\ -\sqrt{\beta_{11}\beta_{22}} & \beta_{22} \end{bmatrix}.$$

Because Π_α and Π_β are linearly independent, we have $\alpha_{11}\beta_{22} - \alpha_{22}\beta_{11} \neq 0$. Thus,

$$\det(a\Pi_\alpha + b\Pi_\beta) = ab\left(\alpha_{11}\beta_{22} + \beta_{11}\alpha_{22} - 2\sqrt{\alpha_{11}\alpha_{22}}\sqrt{\beta_{11}\beta_{22}}\right)$$

is positive if $ab > 0$. Hence, for $a > 0$ and $b > 0$, $a\Pi_\alpha + b\Pi_\beta$ has positive diagonal entries and positive determinant and, thus, is strictly copositive. Therefore, for $a > 0$, $b > 0$, and $c < 0$ sufficiently close to 0, A is not copositive, and $L(A)$ is copositive.

- Suppose that $\beta_{11} = 0$. Then, $\beta_{12} \geq 0$.
  - If $\beta_{12} > 0$, let $a > 0$ and $c < 0$ be so that $a\alpha_{11} + c\gamma_{11} \geq 0$ and $a\alpha_{22} + c\gamma_{22} \geq 0$. Then, for $b > 0$ large enough, A is not copositive, and $L(A) \geq 0$ is copositive.
  - If $\beta_{12} = 0$, then $\beta_{22} > 0$. For $a > 0$, $c < 0$ sufficiently close to 0, and $b > 0$ large enough, A is not copositive, and $L(A)$ is copositive.
- The proof is similar if $\beta_{22} = 0$.

□

Lemma 6.6.8 We have $\alpha_{11} = \alpha_{12} = \beta_{11} = \beta_{22} = \gamma_{22} = \gamma_{12} = 0$ or $\alpha_{22} = \alpha_{12} = \beta_{11} = \beta_{22} = \gamma_{11} = \gamma_{12} = 0$.

Proof By Lemma 6.6.7, $\alpha_{11}\alpha_{22} = 0$ and $\gamma_{11}\gamma_{22} = 0$.

- Suppose that $\alpha_{11} = 0$. We will show that $\alpha_{12} = \beta_{11} = \beta_{22} = \gamma_{22} = \gamma_{12} = 0$, which implies $\alpha_{22}, \beta_{12}, \gamma_{11} > 0$.
  - If $\beta_{11} \neq 0$ then for $b < 0$ and $a, c > 0$ with $b < -c\frac{\gamma_{11}}{\beta_{11}}$ and $a > \frac{b^2}{c}$, A is copositive, and $L(A)$ is not copositive as its 1, 1 entry is negative. So $\beta_{11} = 0$ and, then, $\beta_{12} \geq 0$.
  - Suppose that $\alpha_{12} = 0$ (and $\beta_{11} = 0$).
    - If $\beta_{22} \neq 0$, for $b > 0$, $c = 0$ and $a < 0$ sufficiently close to 0, A is not copositive, and $L(A)$ is copositive. If $\gamma_{22} \neq 0$, then $\gamma_{11} = 0$ and $\gamma_{12} \geq 0$. Thus, for $c > 0$, $b = 0$ and $a < 0$ sufficiently small, A is not copositive, and $L(A)$ is copositive. Thus, $\beta_{22} = \gamma_{22} = 0$.
    - If $\alpha_{11} = \alpha_{12} = \beta_{22} = \gamma_{22} = \beta_{11} = 0$ and $\gamma_{12} \neq 0$, then $\gamma_{12} > 0$. For $a = 0$, $c > 0$ large and $b < 0$ sufficiently small, A is not copositive, and $L(A)$ is copositive. Thus, $\gamma_{12} = 0$.
  - Suppose that $\alpha_{12} \neq 0$ (and $\beta_{11} = 0$). Then $\alpha_{12} > 0$.
    - If $\alpha_{22} \neq 0$ or $\beta_{22} = 0$, for $a > 0$, $c = 0$, and $b < 0$ sufficiently close to 0, A is not copositive and $L(A)$ is copositive.

- If $\alpha_{22} = 0$ and $\beta_{22} \neq 0$, for $b < 0$ and $a, c > 0$ with $b < -c\frac{\gamma_{22}}{\beta_{22}}$ and $a > \frac{b^2}{c}$, A is copositive, and $L(A)$ is not copositive as its 2, 2 entry is negative.

- With similar arguments, we show that if $\alpha_{22} = 0$ then $\alpha_{12} = \beta_{11} = \beta_{22} = \gamma_{11} = \gamma_{12} = 0$.

□

Theorem 6.6.9 A linear transformation $L\colon \mathbf{S}_2 \to \mathbf{S}_2$ maps $\mathbf{C}_2$ onto $\mathbf{C}_2$ if and only if

$$L\left(\begin{bmatrix} 1 & 0 \\ 0 & 0 \end{bmatrix}\right) = \begin{bmatrix} \alpha_{11} & 0 \\ 0 & 0 \end{bmatrix}, L\left(\begin{bmatrix} 0 & 0 \\ 0 & 1 \end{bmatrix}\right) = \begin{bmatrix} 0 & 0 \\ 0 & \gamma_{22} \end{bmatrix}, \\ L\left(\begin{bmatrix} 0 & 1 \\ 1 & 0 \end{bmatrix}\right) = \begin{bmatrix} 0 & \beta_{12} \\ \beta_{12} & 0 \end{bmatrix}, \tag{6.6.2}$$

or

$$L\left(\begin{bmatrix} 1 & 0 \\ 0 & 0 \end{bmatrix}\right) = \begin{bmatrix} 0 & 0 \\ 0 & \alpha_{11} \end{bmatrix}, L\left(\begin{bmatrix} 0 & 0 \\ 0 & 1 \end{bmatrix}\right) = \begin{bmatrix} \gamma_{22} & 0 \\ 0 & 0 \end{bmatrix}, \\ L\left(\begin{bmatrix} 0 & 1 \\ 1 & 0 \end{bmatrix}\right) = \begin{bmatrix} 0 & \beta_{12} \\ \beta_{12} & 0 \end{bmatrix}, \tag{6.6.3}$$

for some $\alpha_{11} > 0$, $\gamma_{22} > 0$ and $\beta_{12} = \sqrt{\alpha_{11}\gamma_{22}}$.

Proof The sufficiency is obvious. Now we show the necessity. From Lemma 6.6.8 it follows that either (6.6.2) or (6.6.3) holds for some $\alpha_{11} \geq 0$, $\gamma_{22} \geq 0$, and $\beta_{12} \geq 0$. Suppose that (6.6.2) holds. The proof is similar if (6.6.3) holds. Because L is a linear transformation that maps $\mathbf{C}_2$ onto $\mathbf{C}_2$ it follows that $\alpha_{11}\gamma_{22}\beta_{12} \neq 0$. Thus, we just need to see that $\beta_{12} = \sqrt{\alpha_{11}\gamma_{22}}$. Let $a, c > 0$ and $b = -\sqrt{ac}$. Then

$$\begin{bmatrix} a & b \\ b & c \end{bmatrix}$$

is copositive, and

$$L\begin{bmatrix} a & b \\ b & c \end{bmatrix} = \begin{bmatrix} a\alpha_{11} & -\sqrt{ac}\beta_{12} \\ -\sqrt{ac}\beta_{12} & c\gamma_{22} \end{bmatrix}$$

is copositive if and only if $-\sqrt{ac}\beta_{12} \geq -\sqrt{ac\alpha_{11}\gamma_{22}}$, which implies $\beta_{12} \leq \sqrt{\alpha_{11}\gamma_{22}}$.

Suppose that $0 < \beta_{12} < \sqrt{\alpha_{11}\gamma_{22}}$. Let $b = \frac{-\sqrt{ac}\sqrt{\alpha_{11}\gamma_{22}}}{\beta_{12}} < -\sqrt{ac}$. Then,

$$\begin{bmatrix} a\alpha_{11} & b\beta_{12} \\ b\beta_{12} & c\gamma_{22} \end{bmatrix}$$

is copositive, and

$$\begin{bmatrix} a & b \\ b & c \end{bmatrix}$$

is not copositive. Thus, we conclude that $\beta_{12} = \sqrt{\alpha_{11}\gamma_{22}}$. □

Corollary 6.6.10 If the linear transformation $L\colon \mathbf{S}_2 \to \mathbf{S}_2$ maps $\mathbf{C}_2$ onto $\mathbf{C}_2$, then there is a monomial matrix $E \in M_n(\mathbb{R})$ such that $L(A) = EAE^T$ for any $A \in \mathbf{S}_2$.

References

[AH08] M. Alanelli and A. Hadjidimos, A new iterative criterion for H-matrices: the reducible case, *Linear Algebra Appl.*, 428 (2008), 2761–2777.

[Al-No88] A. Al-Nowaihi, P-matrices: an equivalent characterization, *J. Algebra*, 112 (1988), 385–387.

[ACE95] L. Andersson, G. Z. Chang, and T. Elfving, Criteria for copositive matrices using simplices and barycentric coordinates, Proceedings of the Workshop on Nonnegative Matrices, Applications and Generalizations and the Eighth Haifa Matrix Theory Conference (Haifa, 1993)], *Linear Algebra Appl.*, 220 (1995), 9–30.

[And80] T. Ando, Inequalities for M-matrices, *Linear Multilinear Algebra*, 8 (1979/80), 291–316.

[Bal70] C. S. Ballantine, Stabilization by a diagonal matrix, *Proceedings of the American Mathematical Society*, 25 (1970), 728–734.

[BaJoh76] C. S. Ballantine and C. R. Johnson, Accretive matrix products, *Linear Multilinear Algebra*, 3 (1975/76), no. 3, 169–185.

[BCN05] R. B. Bapat, M. Catral, and M. Neumann, On functions that preserve M-matrices and inverse M-matrices, *Linear Multilinear Algebra*, 53 (2005), 193–201.

[Bas68] V. J. D. Baston, Extreme copositive quadratic forms, *Acta Arith.*, 15 (1968/1969), 319–327.

[Bau66] L. D. Baumert, Extreme copositive quadratic forms, *Pacific J. Math.*, 19 (1966), 197–204.

[Bau67] L. D. Baumert, Extreme copositive quadratic forms, II, *Pacific J. Math.*, 20 (1967), 1–20.

[B-IG03] A. Ben-Israel and T. N. E. Greville, *Generalized Inverses: Theory and Applications*, 2nd edition, CMS Books in Mathematics/Ouvrages de Mathématiques de la SMC, 15, New York: Springer-Verlag, 2003.

[BCEM12] E. Bendito, A. Carmona, A. M. Encinas, and M. Mitjana, The M-matrix inverse problem for singular and symmetric Jacobi matrices, *Linear Algebra Appl.*, 436 (2012), 1090–1098.

[Ber81] A. Berman, Matrices and the linear complementarity problem, *Linear Algebra Appl.*, 40 (1981), 249–256.

[BDS15] A. Berman, M. Dür, and N. Shaked-Monderer, Open problems in the theory of completely positive and copositive matrices, *Electron. J. Linear Algebra*, 29 (2015), 46–58.

[BH83] A. Berman and D. Hershkowitz, Matrix diagonal stability and its implications, *SIAM J. Algebraic Discrete Methods*, 4 (1983), 377–382.

[BHJ85] A. Berman, D. Hershkowitz, and C. R. Johnson, Linear transformations that preserve certain positivity classes of matrices, *Linear Algebra Appl.*, 68 (1985), 9–29.

[BP94] A. Berman and R. J. Plemmons, *Nonnegative Matrices in the Mathematical Sciences*, SIAM, Philadelphia, 1994.

[BS-M03] A. Berman and N. Shaked-Monderer, *Completely Positive Matrices*, Singapore: World Scientific, 2003.

[BVW78] A. Berman, R. S. Varga, and R. C. Ward, ALPS: matrices with nonpositive off-diagonal entries, *Linear Algebra Appl.*, 21 (1978), 233–244.

[BW77] A. Berman and R. C. Ward, Stability and semipositivity of real matrices, *Bull. Amer. Math. Soc.*, 83 (1977), 262–263.

[BW78] A. Berman and R. C. Ward, ALPS: classes of stable and semipositive matrices, *Linear Algebra Appl.*, 21 (1978), 163–174.

[Bha07] R. Bhatia, *Positive Definite Matrices*, Princeton, NJ: Princeton University Press, 2007.

[BG84] S. Bialas and J. Garloff, Intervals of P-matrices and related matrices, *Linear Algebra Appl.*, 58 (1984), 33–41.

[Blatt62] J. W. Blattner, Bordered matrices, *J. Soc. Indust. Appl. Math.*, 10 (1962) 528–536.

[Bom87] I. M. Bomze, Remarks on the recursive structure of copositivity, *J. Inform. Optim. Sci.*, 8 (1987), 243–260.

[Bom96] I. M. Bomze, Block pivoting and shortcut strategies for detecting copositivity, *Linear Algebra Appl.*, 248 (1996), 161–184.

[Bom00] I. M. Bomze, Linear-time copositivity detection for tridiagonal matrices and extension to block-tridiagonality, *SIAM J. Matrix Anal. Appl.*, 21 (2000), 840–848.

[Bom08] I. M. Bomze, Perron–Frobenius property of copositive matrices, and a block copositivity criterion, *Linear Algebra Appl.*, 429 (2008), 68–71.

[BEFB94] S. Boyd, L. El Ghaoui, E. Feron, and V. Balakrishnan, *Linear Matrix Inequalities in System and Control Theory*, Philadelphia: SIAM, 1994.

[BPS05] R. Bru, F. Pedroche, and D. Szyld, Subdirect sums of nonsingular M-matrices and of their inverses, *Electron. J. Linear Algebra*, 13 (2005), 162–174.

[Car67] D. Carlson, Weakly sign-symmetric matrices and some determinantal inequalities, *Colloq. Math.*, 17 (1967), 123–129.

[CM] D. Carlson and T. Markham, Products of inverse M-matrices, working paper.

[CS94] G. Chang and T. W. Sederberg, Nonnegative quadratic Bézier triangular patches, *Comput. Aided Geom. Design*, 11 (1994), 113–116.

[CLW12] F. Chen, Y. Li, and D. Wang, A new eigenvalue bound for the Hadamard product of an M-matrix and an inverse M-matrix, *Electron. J. Linear Algebra*, 23 (2012), 287–294.

[Che04a] S. Chen, A property concerning the Hadamard powers of inverse M-matrices, *Linear Algebra Appl.*, 381 (2004), 53–60.

[Che04b] S. Chen, A lower bound for the minimum eigenvalue of the Hadamard product of matrices, *Linear Algebra Appl.*, 378 (2004), 159–166.

[Che07a] S. Chen, Proof of a conjecture concerning the Hadamard powers of inverse M-matrices, *Linear Algebra Appl.*, 422 (2007), 477–481.

[Che07b] S. Chen, Inequalities for M-matrices and inverse M-matrices, *Linear Algebra Appl.*, 426 (2007), 610–618.

[CSY01] X. Chen, Y. Shogenji, and M. Yamasaki, Verification for existence of solutions of linear complementarity problems, *Linear Algebra Appl.*, 324 (2001), 15–26.

[CKL92] S. J. Cho, S. Kye, and S. G. Lee, Generalized Choi maps in three-dimensional matrix algebra, *Linear Algebra Appl.*, 171 (1992), 213–224.

[Choi75] M.-D. Choi, Positive semidefinite biquadratic forms, *Linear Algebra Appl.*, 12 (1975), 95–100.

[CKS18a] P. N. Choudhury, R. Kannan, and K. C. Sivakumar, A note on linear preservers of semipositive and minimally semipositive matrices, *Electron. J. Linear Alg.*, 34 (2018), 687–694.

[Cott10] R. W. Cottle, A field guide to the matrix classes found in the literature of the linear complementarity problem, *J. Glob. Optim.*, 46 (2010), 571–580.

[CHL70a] R. W. Cottle, G. J. Habetler, and C. E. Lemke, Quadratic forms semi-definite over convex cones, In *Proceedings of the Princeton Symposium on Mathematical Programming*, Princeton, NJ: Princeton University Press, 1970, 551–565.

[CHL70b] R. W. Cottle, G. J. Habetler, and C. E. Lemke, On classes of copositive matrices, *Linear Algebra Appl.*, 3 (1970), 295–310.

[CPS92] R. W. Cottle, J.-S. Pang, and R. E. Stone, *The Linear Complementarity Problem*, Boston: Academic Press, 1992.

[Cox94] G. E. Coxson, The P-matrix problem is co-NP-complete, *Mathematical Programming*, 64 (1994), 173–178.

[Cox99] G. E. Coxson, Computing exact bounds on elements of an inverse interval matrix is NP-hard, *Reliable Computing*, 5 (1999), 137–142.

[CH69] D. Crabtree and E. V. Haynsworth, An identity for the Schur complement of a matrix, *Proc. AMS*, 22 (1969), 364–366.

[CFJ01] A. Crans, S. Fallat, and C. R. Johnson, The Hadamard core of the totally nonnegative matrices, *Linear Algebra Appl.*, 328 (2001), 203–222.

[Cry76] C. W. Cryer. Some properties of totally positive matrices, *Linear Algebra Appl.*, 15 (1976), 1–25.

[Dan90] G. Danninger, A recursive algorithm for determining (strict) copositivity of a symmetric matrix, Proceedings of the 14th Symposium on Operations Research (Ulm, 1989), *Methods Oper. Res.*, 62, Frankfurt am Main: Hain 45–52.

[DH00] L. M. DeAlba and L. Hogben, Completions of P-matrix patterns, *Linear Algebra Appl.*, 319 (2000), 83–102.

[DMS09] C. Dellacherie, S. Martinez, and J. San Martin, Hadamard functions of inverse M-matrices, *SIAM J. Matrix Anal. Appl.*, 31 (2009), 289–315.

[DMS13] C. Dellacherie, S. Martinez, and J. San Martin, The class of inverse M-matrices associated to random walks, *SIAM J. Matrix Anal. Appl.*, 34 (2013), 831–854.

[Dia62] P. H. Diananda, On nonnegative forms in real variables some or all of which are nonnegative, *Proc. Cambridge Philos. Soc.*, 58 (1962), 17–25.

[Dic10] P. J. C. Dickinson, An improved characterization of the interior of the completely positive cone, *Electron. J. Linear Algebra*, 20 (2010), 723–729.

[Dic11] P. J. C. Dickinson, Geometry of the copositive and completely positive cones, *J. Math. Anal. Appl.*, 380 (2011), 377–395.

[DDGH13a] P. J. C. Dickinson, M. Dür, L. Gijben, and R. Hildebrand, Irreducible elements of the copositive cone, *Linear Algebra Appl.*, 439 (2013), no. 6, 1605–1626.

[DDGH13b] P. J. C. Dickinson, M. Dür, L. Gijben, and R. Hildebrand, Scaling relationship between the copositive cone and Parrilo's first level approximation, *Optim. Lett.*, 7 (2013), no. 8, 1669–1679.

[DH16] P. J. C. Dickinson and R. Hildebrand, Considering copositivity locally, *J. Math. Anal. Appl.*, 437 (2016), no. 2, 1184–1195.

[DLQ18] W. Ding, Z. Luo, and L. Qi, P-tensors, P_0-tensors, and their applications, *Linear Algebra Appl.*, 555 (2018), 336–354.

[DGJJT16] J. Dorsey, T. Gannon, N. Jacobson, C. R. Johnson, and M. Turnansky, Linear preservers of semi-positive matrices, *Linear Multilinear Algebra*, 64 (2016), no. 9, 1853–1862.

[DGJT16] J. Dorsey, T. Gannon, C. R. Johnson, and M. Turnansky, New results about semipositive matrices, *Czechoslovak Math. Journal*, 66 (2016), no. 3, 621–632.

[Dur10] M. Dür, Copositive Programming – A Survey, In M. Diehl, F. Glineur, Elias Jarlebring, and W. Michiels, Editors, *Recent Advances in Optimization and Its Applications in Engineering*, pp. 3–20, New York: Springer, 2010.

[E] G. M. Engel, personal communication.

[Gha90] L. El Ghaoui, *Robustness of Linear Systems to Parameter Variations*, Ph.D. Dissertation, Stanford University, 1990.

[EMS02] L. Elsner, V. Monov, and T. Szulc, On some properties of convex matrix sets characterized by P-matrices and block P-matrices, *Linear Multilinear Algebra*, 50 (2002), 199–218.

[ENN98] L. Elsner, R. Nabben, and M. Neumann, Orthogonal bases that lead to symmetric nonnegative matrices. *Linear Algebra Appl.*, 271 (1998), 323–343.

[ES98] L. Elsner and T. Szulc, Block P-matrices, *Linear Multilinear Algebra*, 44 (1998), 1–12.

[ES00] L. Elsner and T. Szulc, Convex sets of Schur stable and stable matrices, *Linear Multilinear Algebra*, 48 (2000), 1–19.

[ES76] G. M. Engel and H. Schneider, The Hadamard–Fischer inequality for a class of matrices defined by eigenvalue monotonicity, *Linear Multilinear Algebra*, 4 (1976), 155–176.

[EJ91] C. A. Eschenbach and C. R. Johnson, Sign patterns that require real, nonreal or pure imaginary eigenvalues, *Linear Multilinear Algebra*, 29 (1991), 299–311.

[Fal01] S. M. Fallat, Bidiagonal factorizations of totally nonnegative matrices, *American Mathematical Monthly*, 108 (2001), 697–712.

[FHJ98] S. M. Fallat, H. T. Hall, and C. R. Johnson, Characterization of product inequalities for principal minors of M-matrices and inverse M-matrices, *Quart. J. Math. Oxford Ser.*, (2), 49 (1998), 451–458.

[FJ11] S. M. Fallat and C. R. Johnson, *Totally Nonnegative Matrices*, Princeton, NJ: Princeton University Press, 2011.

[FJSvdD98] S. M. Fallat, C. R. Johnson, R. L. Smith, and P. van den Driessche, Eigenvalue location for nonnegative and Z-matrices. *Linear Algebra Appl.*, 277 (1998), 187–198.

[FJTU00] S. M. Fallat, C. R. Johnson, J. R. Torregrosa, and A. M. Urbano, P-matrix completions under weak symmetry assumptions, *Linear Algebra Appl.*, 312 (2000), 73–91.

[FT02] S. M. Fallat and M. J. Tsatsomeros, On the Cayley transform of certain positivity classes of matrices, *Electron. J. Linear Algebra*, 9 (2002), 190–196.

[Fan60] K. Fan, Note on M-matrices, *Quart. J. Math Oxford Ser.*, (2), 11 (1960), 3–49.

[Fang89] L. Fang, On the spectra of P- and P_0-matrices, *Linear Algebra Appl.*, 119 (1989), 1–25.

[FP1912] M. Fekete and G. Pólya, Über ein problem von Laguerre, *Rendiconti del Circolo Matematico di Palermo*, 34 (1912), 89–100, 110–120.

[Fie76] M. Fiedler, Aggregation in graphs, *Combinatorics*, 18 (1976), 315–330.

[Fie86] M. Fiedler, *Special Matrices and Their Applications in Numerical Mathematics*, Leiden, Netherlands: Martinus Nijhoff Publishers, 1986.

[Fie98a] M. Fiedler, Ultrametric matrices in Euclidean point spaces, *Electron. J. Linear Algebra*, 3 (1998), 23–30.

[Fie98b] M. Fiedler, Some characterizations of symmetric inverse M-matrices, *Linear Algebra Appl.*, 275/276 (1998), 179–187.

[Fie00] M. Fiedler, Special ultrametric matrices and graphs, *SIAM J. Matrix Anal. Appl.*, 22 (2000), 106–113.

[FJM87] M. Fiedler, C. R. Johnson, and T. L. Markham, Notes on inverse M-matrices, *Linear Algebra Appl.*, 91 (1987), 75–81.

[FJMN85] M. Fiedler, C. R. Johnson, T. L. Markham, and M. Neumann, A trace inequality for M-matrices and the symmetrizability of a real matrix by a positive diagonal matrix, *Linear Algebra Appl.*, 71 (1985), 81–94.

[FM88a] M. Fiedler and T. L. Markham, An inequality for the Hadamard product of an M-matrix and an inverse M-matrix, *Linear Algebra Appl.*, 101 (1988), 1–8.

[FM88b] M. Fiedler and T. L. Markham, A characterization of the closure of inverse M-matrices, *Linear Algebra Appl.*, 105 (1988), 209–223.

[FM90] M. Fiedler and T. L. Markham, Some connections between the Drazin inverse, P-matrices, and the closure of inverse M-matrices, *Linear Algebra Appl.*, 132 (1990), 163–172.

[FM92] M. Fiedler and T. L. Markham, A classification of matrices of class Z, *Linear Algebra Appl.*, 173 (1992), 115–124.

[FMN83] M. Fiedler, T. L. Markham, and M. Neumann, Classes of products of M-matrices and inverse M-matrices, *Linear Algebra Appl.*, 52/53 (1983), 265–287.

[FP62] M. Fiedler and V. Pták, On matrices with non-positive off-diagonal elements and positive principal minors, *Czech. Math. J.*, 12 (1962), 382–400.

[FP66] M. Fiedler and V. Pták, Some generalizations of positive definiteness and monotonicity, *Numerische Mathematik*, 9 (1966), 163–172.

[FP67] M. Fiedler and V. Pták, Diagonally dominant matrices, *Czech. Math. J.*, 17 (1967), 420–433.

[FF58] M. E. Fischer and A. T. Fuller, On the stabilization of matrices and the convergence of linear iterative processes, *Proc. Cambridge Philos. Soc.*, 54 (1958), 417–425.

[FHS87] S. Friedland, D. Hershkowitz, and H. Schneider, Matrices whose powers are M-matrices or Z-matrices, *Transactions of the American Mathematical Society*, 300 (1987), 343–366.

[FJZ19] S. Furtado, C. R. Johnson, and Y. Zhang, Linear preservers of copositive matrices, Preprint, 2019.

[Gad58] J. W. Gaddum, Linear inequalities and quadratic forms, *Pacific J. Math.*, 8 (1958), 411–414.

[GN65] D. Gale and H. Nikaido, The Jacobian matrix and global univalence of mappings, *Math Ann.*, 159 (1965), 81–93.

[Gan59] F. R. Gantmacher, *The Theory of Matrices*, Vol. I, New York: Chelsea, 1959.

[GK1935] F. R. Gantmacher and M. G. Krein, Sur les matrices oscillatoires, *Comptes Rendus Mathématique Académie des Sciences Paris*, 201 (1935), 577–579.

[GZ79] C. B. Garcia and W. I. Zangwill, On univalence and P-matrices, *Linear Algebra Appl.*, 24 (1979), 239–250.

[Gar] J. Garloff, private communication.

[Gar82] J. Garloff, Criteria for Sign Regularity of Sets of matrices, *Linear Algebra Appl.*, 44 (1982), 153–160.

[Gar96] J. Garloff, Vertex Implications for Totally Nonnegative Matrices, in *Total Positivity and Its Applications*, M. Gasca and C. A. Micchelli, Editors, Boston: Kluwer Academic, 1996, 103–107.

[GP96] M. Gasca and J. M. Peña, On factorizations of totally positive matrices, In M. Gasca and C. A. Micchelli, Editors, *Total Positivity and Its Applications* (Mathematics and Its Applications), Boston: Kluwer Academic, 359, 1996, 109–130.

[GG19] I. B. Gharbia and J. C. Gilbert, An algorithmic characterization of P-matricity II: Adjustments, refinements, and validation, *SIAM J. Matrix Anal. Appl.*, 40 (2019), 800–813.

[GS14] F. Goldberg and N. Shaked-Monderer, On the maximal angle between copositive matrices, *Electron. J. Linear Algebra*, 27 (2014), 837–850.

[Gol80] M. C. Golumbic, *Algorithmic Graph Theory and Perfect Graphs*, New York: Academic Press, 1980.

[Gow89] M. S. Gowda, Pseudomonotone and copositive star matrices, *Linear Algebra Appl.*, 113 (1989), 107–118.

[Gow12] M. S. Gowda, On copositive and completely positive cones, and **Z**-transformations, *Electron. J. Linear Algebra*, 23 (2012), 198–211.

[GR00] M. S. Gowda and G. Ravindran, Algebraic univalence theorems for nonsmooth functions, *J. Math. Anal. Appl.*, 252 (2000), 917–935.

[GS06] M. S. Gowda and R. Sznajder, Automorphism invariance of P- and GUS-properties of linear transformations on Euclidean Jordan algebras, *Mathematics of Operations Research*, 31 (2006), 109–123.

[GST04] M. S. Gowda, R. Sznajder, and J. Tao, Some P-properties for linear transformations on Euclidean Jordan algebras, *Linear Algebra Appl.*, 393 (2004), 203–232.

[GTR12] M. S. Gowda, J. Tao, and G. Ravindran, On the P-property of Z and Lyapunov-like transformations on Euclidean Jordan algebras, *Linear Algebra Appl.*, 436 (2012), 2201–2209.

[GT06a] K. Griffin and M. Tsatsomeros, Principal minors, Part I: A method for computing all the principal minors of a matrix, *Linear Algebra Appl.*, 419 (2006), 107–124.

[GT06b] K. Griffin and M. Tsatsomeros, Principal minors, Part II: The principal minor assignment problem, *Linear Algebra Appl.*, 419 (2006), 125–171.

[GJSW84] R. Grone, C. R. Johnson, E. M. de Sa, and H. Wolkowicz, Positive definite completions of partial Hermitian matrices, *Linear Algebra Appl.*, 58 (1984), 109–124.

[Had83] K. P. Hadeler, On copositive matrices, *Linear Algebra Appl.*, 49 (1983), 79–89.

[Had97] P. Hadjicostas, Copositive matrices and Simpson's paradox, *Linear Algebra Appl.*, 264 (1997), 475–488.

[Hal67] M. Hall, Jr., *Combinatorial Theory*, 2nd edition, Waltham, MA: Blaisdell Publishing Co., 1967.

[HN63] M. Hall, Jr. and M. Newman, Copositive and completely positive quadratic forms, *Proc. Cambridge Philos. Soc.*, 59 (1963), 329–339.

[Hay68] E. Haynsworth, Determination of the inertia of a partitioned Hermitian matrix, *Linear Algebra Appl.*, 1 (1968), 73–81.

[HH69] E. Haynsworth and A. J. Hoffman, Two remarks on copositive matrices, *Linear Algebra Appl.*, 2 (1969), 387–392.

[Hel1923] E. Helly, Über Mengen konvexer Körper mit gemeinschaftlichen Punkte, *Jahresbericht der Deutschen Mathematiker-Vereinigung*, 32 (1923), 175–176.

[Her83] D. Hershkowitz, On the spectra of matrices having nonnegative sums of principal minors, *Linear Algebra Appl.*, 55 (1983), 81–86.

[HB83] D. Hershkowitz and A. Berman, Localization of the spectra of P- and P_0-matrices, *Linear Algebra Appl.*, 52/53 (1983), 383–397.

[HJ86a] D. Hershkowitz and C. R. Johnson, Linear transformations that map the P-matrices into themselves, *Linear Algebra Appl.*, 74 (1986), 23–38.

[HJ86b] D. Hershkowitz and C. R. Johnson, Spectra of matrices with P-matrix powers, *Linear Algebra Appl.*, 80 (1986), 159–171.

[HK03] D. Hershkowitz and N. Keller, Positivity of principal minors, sign symmetry and stability, *Linear Algebra Appl.*, 364 (2003), 105–124.

[HS86] D. Hershkowitz and H. Schneider, Matrices with a sequence of accretive powers, *Israel Journal of Mathematics*, 55 (1986), 327–344.

[Hil12] R. Hildebrand, The extreme rays of the 5×5 copositive cone, *Linear Algebra Appl.*, 437 (2012), no. 7, 1538–1547.

[Hil14] R. Hildebrand, Minimal zeros of copositive matrices, *Linear Algebra Appl.*, 459 (2014), 154–174.

[HS10] J. B. Hiriart-Urruty and A. Seeger, A variational approach to copositive matrices, *SIAM Rev.*, 52 (2010), 593–629.

[HP73] A. J. Hoffman and F. Pereira, On copositive matrices with -1, 0, 1 entries, *J. Combinatorial Theory Ser. A*, 14 (1973), 302–309.

[Hog98a] L. Hogben, Completions of inverse M-matrix patterns, *Linear Algebra Appl.*, 282 (1998), 145–160.

[Hog98b] L. Hogben, Completions of M-matrix patterns, *Linear Algebra Appl.*, 285 (1998), 143–152.

[Hndbk06] L. Hogben (Ed.), *Handbook of Linear Algebra*, Boca Raton, FL: CRC Press Inc., 2006.

[Hog07] L. Hogben, The copositive completion problem: unspecified diagonal entries, *Linear Algebra Appl.*, 420 (2007), no. 1, 160–162.

[HJR05] L. Hogben, C. R. Johnson, and R. Reams, The copositive completion problem, *Linear Algebra Appl.*, 408 (2005), 207–211.

[Hol05] O. Holtz, M-matrices satisfy Newton's inequalities, *Proc. Amer. Math. Soc.*, 133 (2005), 711–717.

[Hor90] R. A. Horn, The Hadamard product, in *Proc. Symp. Appl. Math.*, Amer. Math. Soc., 40 (1990), 87–169.

[HJ85] R. A. Horn and C. R. Johnson, *Matrix Analysis*, New York: Cambridge University Press, 1985.

[HJ91] R. A. Horn and C. R. Johnson, *Topics in Matrix Analysis*, New York: Cambridge University Press, 1991.

[HJ13] R. A. Horn and C. R. Johnson, *Matrix Analysis*, 2nd edition, New York: Cambridge University Press, 2013.

[HLS10] R. Huang, J. Liu, and N.-S. Sze, Characterizations of inverse M-matrices with special zero patterns, *Linear Algebra Appl.*, 433 (2010), 994–1000.

[Ikr02] K. D. Ikramov, An algorithm, linear with respect to time, for verifying the copositivity of an acyclic matrix, *Comput. Math. Math. Phys.*, 42 (2002), 1701–1703.

[IS00] K. D. Ikramov and N. V. Savel'eva, Conditionally definite matrices, *J. Math, Sci.*, 98 (2000), 1–50.

[Ima83] I. N. Imam, Tridiagonal and upper triangular inverse M-matrices, *Linear Algebra Appl.*, 55 (1983), 93–104.

[Ima84] I. N. Imam, The Schur complement and the inverse M-matrix problem, *Linear Algebra Appl.*, 62 (1984), 235–240.

[JJP98] G. James, C. R. Johnson, and S. Pierce, Generalized matrix function inequalities on M-matrices, *J. London Math. Soc.*, 57 (1998), 562–582.

[JR99] C. Jansson and J. Rohn, An algorithm for checking regularity of interval matrices, *SIAM Journal Matrix Anal. Appl.*, 20 (1999), 756–776.

[Jar13] B. Jargalsaikhan, Indefinite copositive matrices with exactly one positive eigenvalue or exactly one negative eigenvalue, *Electron. J. Linear Algebra*, 26 (2013), 754–761.

[Joh70] C. R. Johnson, Positive definite matrices, *American Mathematical Monthly*, 77 (1970), 259–264.

[Joh72] C. R. Johnson, *Matrices whose Hermitian Part is Positive Definite*, PhD thesis, California Institute of Technology (1972).

[Joh73] C. R. Johnson, An inequality for matrices whose symmetric part is positive definite, *Linear Algebra Appl.*, 6 (1973), 13–18.

[Joh75a] C. R. Johnson, Inequalities for a complex matrix whose real part is positive definite, *Trans. AMS*, 212 (1975), 149–154.

[Joh75b] C. R. Johnson, Hadamard's inequality for matrices with positive definite Hermitian component, *Michigan Mathematical Journal*, 22 (1975), no. 3, 225–228.

[Joh75c] C. R. Johnson, Powers of matrices with positive definite real part, *Proc. AMS*, 50 (1975), 85–91.

[Joh77] C. R. Johnson, A Hadamard product involving M-matrices, *Linear Multilinear Algebra*, 4 (1976/77), 261–264.

[Joh78] C. R. Johnson, Partitioned and Hadamard product matrix inequalities, *J. Res. Nat. Bur. Standards*, 83 (1978), 585–591.

[Joh82] C. R. Johnson, Inverse M-matrices, *Linear Algebra Appl.*, 47 (1982), 195–216.

[Joh83] C. R. Johnson, Sign patterns of inverse nonnegative matrices, *Linear Algebra Appl.*, 55 (1983), 69–80.

[Joh87] C. R. Johnson, Closure properties of certain positivity classes of matrices under various algebraic operations, *Linear Algebra Appl.*, 97 (1987), 243–247.

[Joh90] C. R. Johnson, Matrix completion problems: A survey, In *Proc. Symp. Appl. Math.*, Amer. Math. Soc. 40 (1990), 171–198.

[Joh98] C. R. Johnson, Olga, matrix theory and the Taussky unification problem, *Linear Algebra Appl.*, 280 (1998), 39–49.

[JKS94] C. R. Johnson, M. K. Kerr, and D. P. Stanford, Semipositivity of matrices, *Linear Multilinear Algebra*, 37 (1994), 265–271.

[JK96] C. R. Johnson and B. K. Kroschel, The combinatorially symmetric P-matrix completion problem, *Electron. J. Linear Algebra*, 1 (1996), 59–63.

[JL92] C. R. Johnson, M. Lundquist, Operator matrices with chordal inverse patterns, *Operator Theory: Adv. Appl.*, 59 (1992), 234–251.

[JMP09] C. R. Johnson, C. Marijuan, and M. Pisonero, Matrices and spectra satisfying the Newton inequalities, *Linear Algebra Appl.*, 430 (2009), 3030–3046.

[JMS95] C. R. Johnson, W. D. McCuaig, and D. P. Stanford, Sign patterns that allow minimal semipositivity, Special issue honoring Miroslav Fiedler and Vlastimil Ptak, *Linear Algebra Appl.*, 223/224 (1995), 363–373.

[JN13] C. R. Johnson and S. K. Narayan, When the positivity of the leading principal minors implies the positivity of all principal minors of a matrix, *Linear Algebra Appl.*, 439(2013), 2934–2947.

[JNT96] C. R. Johnson, M. Neumann, and M. Tsatsomeros, Conditions for the positivity of determinants, *Linear Multilinear Algebra*, 40 (1996), 241–248.

[JO05] C. R. Johnson and D. D. Olesky, Rectangular submatrices of inverse M-matrices and the decomposition of a positive matrix as a sum, *Linear Algebra Appl.*, 409 (2005), 87–99.

[JOTvdD93] C. R. Johnson, D. D. Olesky, M. Tsatsomeros, and P. van den Driessche, Spectra with positive elementary symmetric functions, *Linear Algebra Appl.*, 180 (1993), 247–261.

[JOvdD03] C. R. Johnson, D. D. Olesky, and P. van den Driessche, Matrix classes that generate all matrices with positive determinant, *SIAM J. Matrix Anal. Appl.*, 25 (2003), 285–294.

[JOvdD95] C. R. Johnson, D. D. Olesky, and P. van den Driessche, Sign determinancy in LU factorization of P-matrices, *Linear Algebra Appl.*, 217 (1995), 155–166.

[JOvdD01] C. R. Johnson, D. D. Olesky, and P. van den Driessche, Successively ordered elementary bidiagonal factorization, *SIAM J. Matrix Anal. Appl.* 22 (2001), 1079–1088.

[JOvdD03] C. R. Johnson, D. D. Olesky, and P. van den Driessche, Matrix classes that generate all matrices with positive determinant, *SIAM J. Matrix Anal. Appl.*, 25 (2003), 285–294.

[JR05] C. R. Johnson and R. Reams, Spectral theory of copositive matrices, *Linear Algebra Appl.*, 395 (2005), 275–281.

[JR08] C. R. Johnson and R. Reams, Constructing copositive matrices from interior matrices, *Electron. J. Linear Algebra*, 17 (2008), 9–20.

[JS96] C. R. Johnson and R. L. Smith, The completion problem for M-matrices and inverse M-matrices, *Linear Algebra Appl.*, 241–243 (1996), 655–667.

[JS99] C. R. Johnson and R. L. Smith, The symmetric inverse M-matrix completion problem, *Linear Algebra Appl.*, 290 (1999), 193–212.

[JS01a] C. R. Johnson and R. L. Smith, Path product matrices, *Linear Multilinear Algebra*, 46 (1999), 177–191.

[JS01b] C. R. Johnson and R. L. Smith, Almost principal minors of inverse M-matrices, *Linear Algebra Appl.*, 337 (2001), 253–265.

[JS01c] C. R. Johnson and R. L. Smith, Linear interpolation problems for matrix classes and a transformational characterization of M-matrices, *Linear Algebra Appl.*, 330 (2001), 43–48.

[JS02] C. R. Johnson and R. L. Smith, Intervals of inverse M-matrices, *Reliab. Comput.*, 8 (2002), 239–243.

[JS07a] C. R. Johnson and R. L. Smith, Positive, path product, and inverse M-matrices, *Linear Algebra Appl.*, 421 (2007), 328–337.

[JS07b] C. R. Johnson and R. L. Smith, Path product matrices and eventually inverse M-matrices, *SIAM J. Matrix Anal. Appl.*, 29 (2007), 370–376.

[JS11] C. R. Johnson and R. L. Smith, Inverse M-matrices, II, *Linear Algebra Appl.*, 435 (2011), 953–983.

[JS93] C. R. Johnson and D. P. Stanford, Qualitative semipositivity, Combinatorial and graph-theoretical problems in linear algebra, *IMA Vol. Math. Appl.*, 50 (1993), 99–105.

[JT11] C. R. Johnson and Z. Tong, Equilibrants, semipositive matrices, calculation and scaling, *Linear Algebra Appl.*, 434 (2011), 1638–1647.

[JT95] C. R. Johnson and M. J. Tsatsomeros, Convex sets of nonsingular and P-matrices, *Linear Multilinear Algebra*, 38 (1995), 233–239.

[JX93] C. R. Johnson and C. Xenophontos, Irreducibilty and primitivity of Perron complements: applications of the compressed directed graph, in Graph Theory and Sparse Matrix Computation, *IMA Vol. Math. Appl.*, 56 (1993), 101–106.

[JTU02] C. Jordán, J. R. Torregrosa, and A. M. Urbano, Inverse M-matrix completion problem with zeros in the inverse completion, *Appl. Math. Lett.*, 15 (2002), 677–684.

[KS14] M. Rajesh Kannan and K. C. Sivakumar, $P_{\dagger}$-matrices: a generalization of *P*-matrices, *Linear Multilinear Algebra*, 62 (2014), 1–12.

[Kap00] W. Kaplan, A test for copositive matrices, *Linear Algebra Appl.*, 313 (2000), 203–206.

[Kap01] W. Kaplan, A copositivity probe, *Linear Algebra Appl.*, 337 (2001), 237–251.

[Kel69] E. L. Keller, *Quadratic Optimization and Linear Complementarity*, Ph.D. thesis, University of Michigan, Ann Arbor, 1969.

[Kel72] R. Kellogg, On complex eigenvalues of M and P matrices, *Numer. Math.*, 19 (1972), 170–175.

[KB85] O. Kessler and A. Berman, Matrices with a transitive graph and inverse M-matrices, *Linear Algebra Appl.*, 71 (1985), 175–185.

[KOSvdD06] I. J. Kim, D. D. Olesky, B. L. Shader, and P. van den Driessche, Sign patterns that allow a positive or nonnegative left Inverse, *SIAM J. Matrix Anal. Appl.*, 29 (2007), 554–565.

[KN91] I. Koltracht and M. Neumann, On the inverse M-matrix problem for real symmetric positive-definite Toeplitz matrices, *SIAM J. Matrix Anal. Appl.*, 12 (1991), 310–320.

[Kus16] V. Y. Kushel, On the positive stability of P^2-matrices, *Linear Algebra Appl.*, 503(2016), 190–214.

[Lan90] H. Langer, Strictly copositive matrices and ESS's, *Arch. Math.* (Basel), 55 (1990), 516–520.

[Las14] J. B. Lasserre, New approximations for the cone of copositive matrices and its dual, *Math. Program. Ser. A*, 144 (2014), 265–276.

[Lew80] M. Lewin, Totally nonnegative, M-, and Jacobi matrices, *SIAM J. Alg. Disc. Meth.*, 1 (1980), 419–421.

[Lew89] M. Lewin, On inverse M-matrices, *Linear Algebra Appl.*, 118 (1989), 83–94.

[LN80] M. Lewin and M. Neumann, On the inverse M-matrix problem for (0, 1)-matrices, *Linear Algebra Appl.*, 30 (1980), 41–50.

[LLHNT98] B. Li, L. Li, M. Harada, H. Niki, and M. Tsatsomeros, An iterative criterion for H-matrices, *Linear Algebra Appl.*, 271 (1998), 179–190.

[LCW09] Y. T. Li, F. B. Chen, and D. F. Wang, New lower bounds on eigenvalue of the Hadamard product of an M-matrix and its inverse, *Linear Algebra Appl.*, 430 (2009), 1423–1431.

[Loe89] R. Loewy, Linear transformations which preserve or decrease rank, *Linear Algebra Appl.*, 121 (1989), 151–161.

[Man69] O. L. Mangasarian, *Nonlinear Programming*, Corrected reprint of the 1969 original, Classics in Applied Mathematics, 10, Philadelphia: SIAM, 1994.

[Man78] O. L. Mangasarian, Characterization of linear complementarity problems as linear problems. *Mathematical Programming Study*, 7 (1978), 74–87.

[MM59] M. Marcus and B. N. Moyls, Linear transformations on algebras of matrices, *Canadian J. Math*, 11 (1959), 61–66.

[Mark72] T. L. Markham, Nonnegative matrices whose inverses are M-matrices, *Proc. Amer. Math. Soc.*, 36 (1972), 326–330.

[Mark79] T. L. Markham, Two properties of M-matrices, *Linear Algebra Appl.*, 28 (1979), 131–134.

[MS98] T. L. Markham and R. L. Smith, A Schur complement inequality for certain P-matrices, *Linear Algebra Appl.*, 281 (1998), 33–41.

[MJ81] D. H. Martin and D. H. Jacobson, Copositive matrices and definiteness of quadratic forms subject to homogeneous linear inequality constraints, *Linear Algebra Appl.*, 35 (1981), 227–258.

[MMS94] S. Martínez, G. Michon, and J. San Martín, Inverses of strictly ultrametric matrices are of Stieltjes type, *SIAM J. Matrix Anal. Appl.*, 15 (1994), 98–106.

[MSZ03] S. Martínez, J. San Martín, and X.-D. Zhang, A new class of inverse M-matrices of tree-like type, *SIAM J. Matrix Anal. Appl.*, 24 (2003), 1136–1148.

[MH06] J. H. Mathews and R. W. Howell, *Complex Analysis for Mathematics and Engineering*, 5th edition, Sudbury, MA: Jones and Bartlett, 2006.

[MP90] R. Mathias and J.-S. Pang, Error bounds for the linear complementarity problem with a P-matrix, *Linear Algebra Appl.*, 132 (1990), 123–136.

[MNST95] J. J. McDonald, M. Neumann, H. Schneider, and M. J. Tsatsomeros, Inverse M-matrix inequalities and generalized ultrametric matrices, *Linear Algebra Appl.*, 220 (1995), 321–341.

[MNST96] J. J. McDonald, M. Neumann, H. Schneider, and M. J. Tsatsomeros, Inverses of unipathic M-matrices, *SIAM J. of Matrix Anal. Appl.*, 17 (1996), 1025–1036.

[Meg88] N. Megiddo, A note on the complexity of P-matrix LCP and computing an equilibrium, Research report RJ 6439, 5 pages, San Jose, CA: IBM Almaden Research Center, 1988.

[MP91] N. Megiddo and C. H. Papadimitriou, On total functions, existence theorems and computational complexity, *Theoretical Computer Science*, 81 (1991), 317–324.

[MT09] C. Mendes-Araújo and J. R. Torregrosa, Sign pattern matrices that admit M-, N-, P- or inverse M-matrices, *Linear Algebra Appl.*, 431 (2009), 724–731.

[Mey89a] C. D. Meyer, Uncoupling the Perron eigenvector problem, *Linear Algebra Appl.*, 114/115 (1989), 69–94.

[Mey89b] C. D. Meyer, Stochastic complementation, uncoupling Markov chains, and the theory of nearly reducible systems, *SIAM Rev.*, 31 (1989), 240–272.

[MND01] S. R. Mohan, S. K. Neogy, and A. K. Das, On the classes of fully copositive and fully semimonotone matrices, *Linear Algebra Appl.*, 323 (2001), 87–97.

[Mor03] W. D. Morris, Recognition of hidden positive row diagonally dominant matrices, *Electron. J. Linear Algebra*, 10 (2003), 102–105.

[MN07] W. D. Morris and M. Namiki, Good hidden P-matrix sandwiches, *Linear Algebra Appl.*, 426 (2007), 325–341.

[Mot52] T. S. Motzkin, Copositive quadratic forms, *National Bureau of Standards*, Report 1818 (1952), 11–12.

[MS65] T. S. Motzkin and E. G. Strauss, Maxima for graphs and a new proof of a theorem of Turán, *Canad. J. Math.*, 17 (1965), 533–540.

[Mur77] B. B. Murphy, On the inverses of M-matrices, *Proc. Amer. Math. Soc.*, 62 (1977), 196–198.

[MP98] G. S. R. Murthy, and T. Parthasarathy, Fully copositive matrices, *Math. Programming Ser. A*, 82 (1998), 401–411.

[Mu71] K. G. Murty, On a characterization of P-matrices, *SIAM J. Appl. Math.*, 20 (1971), 378–384.

[MK87] K. G. Murty and S. N. Kabadi, Some NP-complete problems in quadratic and nonlinear programming, *Math. Programming*, 39 (1987), 117–129.

[Nab00] R. Nabben, A class of inverse M-matrices, *Electron. J. Linear Algebra*, 7 (2000), 53–58.

[NV94] R. Nabben and R. S. Varga, A linear algebra proof that the inverse of a strictly ultrametric matrix is a strictly diagonally dominant Stieltjes matrix, *SIAM J. Matrix Anal. Appl.*, 15 (1994), 107–113.

[NV95] R. Nabben and R. S. Varga, Generalized ultrametric matrices – a class of inverse M-matrices, *Linear Algebra Appl.*, 220 (1995), 365–390.

[Neu98] M. Neumann, A conjecture concerning the Hadamard product of inverses of M-matrices, *Linear Algebra Appl.*, 285 (1998), 277–290.

[Neu00] M. Neumann, Inverses of Perron complements of inverse M-matrices, *Linear Algebra Appl.*, 313 (2000), 163–171.

[Neu98] M. Neumann, A conjecture concerning the Hadamard product of inverses of M-matrices, *Linear Algebra Appl.*, 285 (1998), 277–290.

[NP79] M. Neumann and R. Plemmons, Generalized inverse-positivity and splittings of M-matrices, *Linear Algebra Appl.*, 23 (1979), 21–35.

[NP80] M. Neumann and R. Plemmons, M-matrix characterizations II: General M-matrices, *Linear Multilinear Algebra*, 9 (1980), 211–225.

[NS11] M. Neumann and N.-S. Sze, On the inverse mean first passage matrix problem and the inverse M-matrix problem, *Linear Algebra Appl.*, 434 (2011), 1620–1630.

[Ober] *Nonnegative matrices, M-matrices and their generalizations*, Oberwolfach, November 26–December 2, 2000. D. Hershkowitz, V. Mehrmann, and H. Schneider, organizers.

[Pan79a] J.-S. Pang, Hidden Z-matrices with positive principal minors, *Linear Algebra Appl.*, 23 (1979), 201–215.

[Pan79b] J.-S. Pang, On discovering hidden Z-matrices, In C. V. Coffman and G. Fix, editors, *Constructive Approaches to Mathematical Models*, pp. 231–241, Academic Press, 1979.

[Parth83] T. Parthasarathy, *On Global Univalence Theorems*, Lecture Notes in Math., Vol. 977, New York: Springer-Verlag, 1983.

[Parth96] T. Parthasarathy, Application of P matrices to economics, *Applied Abstract Algebra*, pp. 78–95, Delhi: University of Delhi, 1996.

[Pen95] J. M. Peña, M-matrices whose inverses are totally positive, *Linear Algebra Appl.*, 221 (1995), 189–193.

[Pen01] J. M. Peña, A class of P-matrices with applications to the localization of the eigenvalues of a real matrix, *SIAM J. Matrix Anal. Appl.*, 22 (2001), 1027–1037.

[Pier92] S. Pierce (Ed.), A survey of linear preserver problems, *Linear Multilinear Algebra*, 33 (1992), 1–129.

[PF93] L. Ping and Y. Y. Feng, Criteria for copositive matrices of order four, *Linear Algebra Appl.*, 194 (1993), 109–124.

[Ple77] R. J. Plemmons, M-matrix characterizations. I – Nonsingular M-matrices, *Linear Algebra Appl.*, 18 (1977), 175–188.

[Pok] P. Pokora, Introduction to Linear Preserver Problems, unpublished remarks.

[PB74] G. Poole and T. Boullion, A survey on M-matrices, *SIAM Rev.*, 16 (1974), 419–427.

[Rock97] R. T. Rockafellar, *Convex Analysis*, Princeton, NJ: Princeton University Press, 1997.

[Roh87] J. Rohn, Inverse-positive interval matrices, *Z. Angew. Math. Mech.*, 67 (1987) no. 5, T492–T493.

[Roh89] J. Rohn, Systems of linear interval equations, *Linear Algebra Appl.*, 126 (1989), 39–78.

[Roh91] J. Rohn, A theorem on P-matrices, *Linear Multilinear Algebra*, 30 (1991), 209–211.

[RR96] J. Rohn and G. Rex, Interval P-matrices, *SIAM J. Matrix Anal. Appl.*, 17 (1996), 1020–1024.

[Rum01] S. Rump, Self-validating methods, *Linear Algebra Appl.*, 324 (2001), 3–13.

[Rum03] S. Rump, On P-matrices, *Linear Algebra Appl.*, 363 (2003), 237–250.

[Rus07] L. Y. Rüst, *The P-Matrix Linear Complementarity Problem – Generalizations and Specializations*, Doctor of Sciences Dissertation, Swiss Federal Institute of Technology, Zurich, 2007.

[Schn65] H. Schneider, Positive Operators and an Inertia Theorem, *Numerische Mathematik*, 7 (1965), 11–17.

[S-MBBJS15] N. Shaked-Monderer, A. Berman, I. M. Bomze, F. Jarre, and W. Schachinger, New results on the cp-rank and related properties of co(mpletely) positive matrices, *Linear Multilinear Algebra*, 63 (2015), no. 2, 384–396.

[ST18] K. C. Sivakumar and M. J. Tsatsomeros, Semipositive matrices and their semipositive cones, *Positivity*, 22 (2018), 379–398.

[Smi81] R. L. Smith, M-matrices whose inverses are stochastic, *SIAM J. Algebraic Discrete Methods*, 2 (1981), 259–265.

[Smi86] R. L. Smith, On the spectrum of N_0-matrices, *Linear Algebra Appl.*, 83 (1986), 129–134.

[Smi87] R. L. Smith, On Markham's M-matrix properties, *Linear Algebra Appl.*, 87 (1987), 189–195.

[Smi88] R. L. Smith, Some notes on Z-matrices, *Linear Algebra Appl.*, 106 (1988), 219–231.

[Smi90] R. L. Smith, Bounds on the spectrum of nonnegative matrices and certain Z-matrices, *Linear Algebra Appl.*, 129 (1990), 13–28.

[Smi94] R. L. Smith, On Fan products of Z-matrices, *Linear Multilinear Algebra*, 37 (1994), 297–302.

[Smi95] R. L. Smith, Some results on a partition of Z-matrices, *Linear Algebra Appl.*, 223/224 (1995), 619–629.

[Smi01] R. L. Smith, On characterizing Z-matrices, *Linear Algebra Appl.*, 338 (2001), 99–104.

[SH98] R. L. Smith and S. A. Hu, Inequalities for monotonic pairs of Z-matrices, *Linear Multilinear Algebra*, 44 (1998), 57–65.

[Son00] Y. Z. Song, On an inequality for the Hadamard product of an M-matrix and its inverse, *Linear Algebra Appl.*, 305 (2000), 99–105.

[SGR99] Y. Song, M. S. Gowda, and G. Ravindran, On some properties of P-matrix sets, *Linear Algebra Appl.*, 290 (1999), 237–246.

[Sou83] G. W. Soules, Constructing nonnegative symmetric matrices, *Linear Multilinear Algebra*, 13 (1983), 241–151.

[Sou12] P. Souplet, Liouville-type theorems for elliptic Schrödinger systems associated with copositive matrices, *Netw. Heterog. Media*, 7 (2012), 967–988.

[SRPM82] M. Stadelmaier, N. Rose, G. Poole, and C. Meyer, Nonnegative matrices with power invariant zero patterns, *Linear Algebra Appl.*, 42 (1982), 23–29.

[Stu98] J. L. Stuart, A polynomial time spectral decomposition test for certain classes of inverse M-matrices, *Electron. J. Linear Algebra*, 3 (1998), 129–141.

[Szu90] T. Szulc, A contribution to the theory of P-matrices, *Linear Algebra Appl.*, 139 (1990), 217–224.

[TSOA07] A. K. Tang, A. Simsek, A. Ozdaglar, and D. Acemoglu, On the stability of P-matrices, *Linear Algebra Appl.*, 426 (2007), 22–32.

[TG05] J. Tao and M. S. Gowda, Some P-properties for nonlinear transformations on Euclidean Jordan algebras, *Mathematics of Operations Research*, 30 (2005), 985–1004.

[Tau58] O. Taussky, Research Problem #124, *Bull. Amer. Math. Soc.*, 64 (1958).

[TV99] J. Todd and R. S. Varga, Hardy's inequality and ultrametric matrices, *Linear Algebra Appl.*, 302/303 (1999), 33–43.

[TT18] P. Torres and M. Tsatsomeros, Stability and convex hulls of matrix powers, *Linear Multilinear Algebra*, 66 (2018), 769–775.

[Tsa00] M. Tsatsomeros, Principal pivot transforms: properties and applications, *Linear Algebra Appl.*, 307 (2000), 151–165.

[TL00] M. Tsatsomeros and L. Li, A recursive test for P-matrices, *BIT*, 40 (2000), 410–414.

[Tsa16] M. Tsatsomeros, Geometric mapping properties of semipositive matrices, *Linear Algebra Appl.*, 498 (2016), 349–359.

[TW19] M. Tsatsomeros and M. Wendler, Semimonotone matrices, *Linear Algebra Appl.*, 578 (2019), 207–224.

[TZ19] M. Tsatsomeros and Y. F. Zhang. The fiber of P-matrices: the recursive construction of all matrices with positive principal minors. *Linear and Multilinear Algebra*, submitted, 2019.

[Val86] H. Väliaho, Criteria for copositive matrices, *Linear Algebra Appl.*, 81 (1986), 19–34.

[Val88] H. Väliaho, Testing the definiteness of matrices on polyhedral cones, *Linear Algebra Appl.*, 101 (1988), 135–165.

[Val89a] H. Väliaho, Almost copositive matrices, *Linear Algebra Appl.*, 116 (1989), 121–134.

[Val89b] H. Väliaho, Quadratic-programming criteria for copositive matrices, *Linear Algebra Appl.*, 119 (1989), 163–182.

[Val91] H. Väliaho, A polynomial-time test for M-matrices, *Linear Algebra Appl.*, 153 (1991), 183–192.

[Van72] J. S. Vandergraft, Applications of partial orderings to the study of positive definiteness, monotonicity, and convergence of iterative methods for linear systems, *SIAM J. Numer. Anal.*, 9 (1972), 97–104.

[Vil1938] J. Ville, Sur a theorie generale des jeux ou intervient Vhabilite des joueurs, *Traite du calcul des probability et de ses applications IV*, 2 (1938), 105–113.

[WZZ00] B. Y. Wang, X. Zhang, and F. Zhang, On the Hadamard product of inverse M-matrices, *Linear Algebra Appl.*, 305 (2000), 23–31.

[Wer94] H. J. Werner, Characterizations of minimal semipositivity, *Linear Multilinear Algebra*, 37 (1994), 273–278.

[Wil77] R. A. Willoughby, The inverse M-matrix problem, *Linear Algebra Appl.*, 18 (1977), 75–94.

[XZ02] N. Xiu and J. Zhang, A characteristic quantity of P-matrices, *Appl. Math. Lett.*, 15 (2002), 41–46.

[XY11] J. Xu and Y. Yao, An algorithm for determining copositive matrices, *Linear Algebra Appl.*, 435 (2011), 2784–2792.

[YX04] C. Yang and C. Xu, Properties of Hadamard product of inverse M-matrices, *Numer. Linear Algebra Appl.*, 11 (2004), 343–354.

[YL09] S. Yang and X. Li, Algorithms for determining the copositivity of a given symmetric matrix, *Linear Algebra Appl.*, 430 (2009), no. 2–3, 609–618.

[Yon00] X. Yong, Proof of a conjecture of Fiedler and Markham, *Linear Algebra Appl.*, 320 (2000), 167–171.

[YW99] X. Yong and Z. Wang, On a conjecture of Fiedler and Markham, *Linear Algebra Appl.*, 288 (1999), 259–267.

[Zha05] F. Zhang (Ed.), *The Schur Complement and Its Applications*, New York: Springer, 2005.

[ZZL11] Y. Zhu, C. Y. Zhang, and J. Liu, Path product and inverse M-matrices, *Electron. J. Linear Algebra*, 22 (2011), 644–652.

Index

Printed in the United States
By Bookmasters